Key Concepts in Biochemistry

KEY CONCEPTS IN BIOCHEMISTRY

Abeer Iqbal

www.arclerpress.com

Key Concepts in Biochemistry

Abeer Iqbal

Arcler Press
224 Shoreacres Road
Burlington, ON L7L 2H2
Canada
www.arclerpress.com
Email: orders@arclereducation.com

ISBN: 978-1-77469-381-0 (Hardcover)

Arcler Press publishes wide variety of books and eBooks. For more information about Arcler Press and its products, visit our website at www.arclerpress.com

ABOUT THE AUTHOR

Abeer Iqbal (M.S Microbiology and Molecular Genetics) is a polished writer with strong scientific knowledge and keen interest in biosciences. She has authored books related to her field and have also been engaged in writing medical blogs/content. She has a passion for learning and writing about scientific advancements and her enthusiasm reflects in her work. She is a nature lover and admires the importance and beauty of little things in life.

TABLE OF CONTENTS

PREFACE

Biochemistry is the branch of science that explores the chemical processes within and related to living organisms. It is a laboratory - based science that brings together biology and chemistry. By using chemical knowledge and techniques, biochemists can understand and solve biological problems.

Biochemistry focuses on processes happening at a molecular level. It focuses on what's happening inside our cells, studying components like proteins, lipids, and organelles. It also looks at how cells communicate with each other, for example during growth or fighting illness. Biochemists need to understand how the structure of a molecule relates to its function, allowing them to predict how molecules will interact.

Biochemistry covers a range of scientific disciplines, including genetics, microbiology, forensics, plant science and medicine. Because of its breadth, biochemistry is very important and advances in this field of science over the past 100 years have been staggering. It's a very exciting time to be part of this fascinating area of study.

The term biochemistry is synonymous with two somewhat older terms: physiological chemistry and biological chemistry. Those aspects of biochemistry that deal with the chemistry and function of very large molecules (e.g., proteins and nucleic acids) are often grouped under the term molecular biology. Biochemistry is a young science, having been known under that term only since about 1900. Its origins, however, can be traced much further back; its early history is part of the early history of both physiology and chemistry.

The particularly significant past events in biochemistry have been concerned with placing biological phenomena on firm chemical foundations.

Before chemistry could contribute adequately to medicine and agriculture, however, it had to free itself from immediate practical demands in order to become a pure science. This happened in the period from about 1650 to 1780, starting with the work of Robert Boyle and culminating in that of Antoine-Laurent Lavoisier, the father of modern chemistry. Boyle questioned the basis of the chemical theory of his day and taught that the proper object of chemistry was to determine the composition of substances.

His contemporary John Mayow observed the fundamental analogy between the respiration of an animal and the burning, or oxidation, of organic matter in air. Then, when Lavoisier carried out his fundamental studies on chemical oxidation, grasping the true nature of the process, he also showed, quantitatively, the similarity between chemical oxidation and the respiratory process.

Photosynthesis was another biological phenomenon that captured the attention of the chemists of the late 18th century. The demonstration, through the combined work of Joseph Priestley, Jan Ingenhousz, and Jean Senebier, that photosynthesis is essentially the reverse of respiration, was a milestone in the development of biochemical thought.

I Despite these early fundamental discoveries, rapid progress in biochemistry had to wait until the development of structural organic chemistry, one of the great achievements of 19th-century science. A living organism contains many thousands of different chemical compounds.

The elucidation of the chemical transformations undergone by these compounds within the living cell is a central problem of biochemistry. Clearly, the determination of the molecular structure of the organic substances that are present in the living cells had to precede the study of the cellular mechanisms, whereby these substances are synthesized and degraded.

This book will introduce the readers to the field of biochemistry and the fundamentals of it. It is precisely designed for readers or students with no prior experience related to biochemistry and its application, and touches upon a diversity of fundamental topics. By the end of the book, readers will understand the basic knowledge of biochemistry and the key concepts involved in biochemistry.

ABC Model

In *Arabidopsis thaliana*, a model for identifying floral organs is presented. It has the floral primordium as comprising four whorls whose developmental fate is resolute by the concentric and combinatorial activity of three classes of genes, denoted A, B, and C, which encode transcription factors. It depends on the APETALA2 gene (AP2) whether whorls 1 and 2 (sepals and petals) are formed; class B determines whorls 2 and 3 (petals and stamens) depending upon the PISTILLATA (PI) and APETALA3 (AP3) genes; class C determines whorl 4 (carpels depending upon the AGAMOUS gene (AG). These genes are described as 'homeotic' albeit they encode transcription factors that contain a MADS box in lieu of homeobox domains. Homologs of these genes occur in other plants.

ABCR

abbr. For ATP-binding cassette transportor retina; other name: rim protein. A protein found in the disc membrane of the outer segment of photoreceptor cells of the retina. It consists of 2273 aminndacids and is surmised to function in the transportance of retinoids. Mutations in the ABCR gene, at 1p21-p23, are associated with Stargardt and age-cognate macular dystrophies.

ABC Transporter

A membrane transport protein having the ABC molecular domain, denominated after ATP-binding cassette, characteristic of all members of a sizably voluminous superfamily of membrane transport proteins that hydrolyze ATP and transfer a diverse array of minute molecules across membranes.

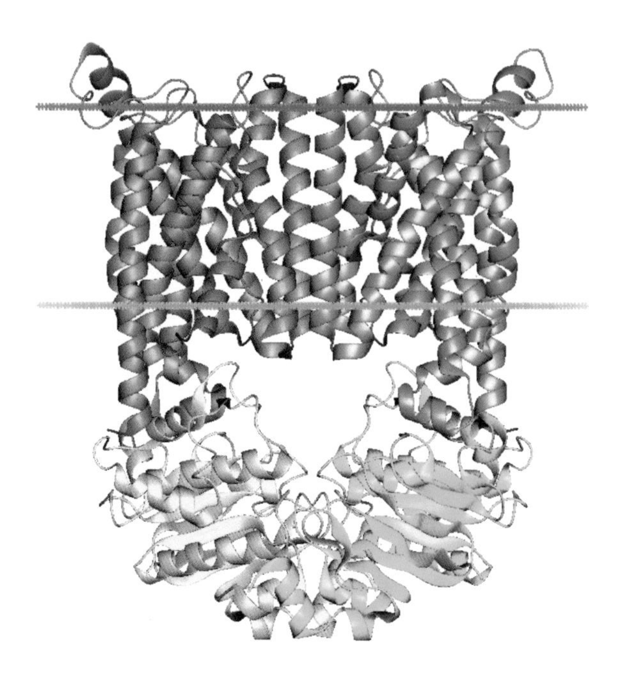

Figure 1. Structure Of The *E. Coli* Btucd Protein, An ABC Transporter That Mediates Vitamin B12 Uptake.

Source : Image By Wikimedia Commons

Abequose

Symbol: Abe; 3,6-dideoxy-D-xylo-hexose; 3,6-dideoxy-D galactose; a deoxysugar that occurs, e.g., in O-categorical chains of lipopolysaccharides in certain serotypes of Salmonella.

Abetalipoproteinemia

Abetalipoproteinaemia is an autosomal recessive disorder in which plasma lipoproteins lack apolipoprotein B. There is defective assembly and secretion both of chylomicrons in intestinal mucosa and of very-low-density lipoproteins in the liver. The cause is a deficiency of the 88 kDa subunit of microsomal triglyceride transfer protein.

ABH Antigens

One of the systems of blood group antigens having determinants associated with oligosaccharide structures. It is the substructure of the ABO system, which was the first human blood group antigen system to be detected, by Austrian-born US pathologist Karl Landsteiner (1868–1943) in 1901, and it remains theamountvital in blood transfusion. Individuals having neither the A nor B antigen express the H antigen, the product of an independent gene belonging to the Hh system. Antigens of the ABH system are oligosaccharide chains, in the erythrocyte carried on band 3 (the anion transporter) and band 4.5 (the glucose transporter), or on ceramide. A highly branched N-glycan, consisting of a trimannosyl-di-N-acetyl-chitobiosyl core with Gal(b1-4) GlcNAc(b1-3) reiterates, forms the substructure of ABH antigens. The H determinant is the precursor; antigen A is composed by additament of N-acetyl-D-galactosamine by fucosylgalactose a-N-acetylgalactosaminyltransferase (EC 2.4.1.40); antigen B is composed by additament of D-galactose by fucosylglycoprotein 3-agalactosyltransferase (EC 2.4.1.37). The terminal sugar residues of consequentiy are:are: H determinant, Fuc(a1-2) Galb-R; A determinant, GalNAc(a1-3) (Fuca1-2) Galb-R; B determinant, Gal(a1-3) (Fuca1-2) Galb-R. The enzyme responsible for integrating the terminal fucosyl residue of the H determinant is galactoside 2-a-L-fucosyltransferase (EC 2.4.1.69).

Abiogenesis

1. The formation of a substance other than by a living organism.
2. Theprinciple that living organisms can emanate from nonliving matter; spontaneous generation.

Ab Initio Gene Prediction

The prognostication of genes in uncharacterized nucleotide sequences utilizing only characteristics of the sequence (codon utilization, compositional partiality, etc.) – that is, without direct reference to other sequences.

Ab Initio Protein Structure Prediction

Presage of the structure of proteins using only properties of amino acid sequence (solvation potentials, secondary structure propensities, etc.) - that is, without direct reference to the structures of kenned homologs.

Abiogenesis Or Spontaneous Generation

The discredited doctrine that living organisms can arise from nonliving materials under current conditions.

Abl

This oncogene is found in the murine Abelson leukemia virus. The human equivalent of this gene is ABL (9q34), which encodes a tyrosine protein kinase. Human ABL activation occurs inopportunely as the result of a reciprocal translocation between chromosomes 9 and 22 in which ABL is linked to the breakpoint cluster region (bCR) of ph1 on chromosome 22(q11), leading to an altered chromosome 22, known as the Philadelphia chromosome (Ph1). The protein product of the spliced genes in the Ph1 chromosome has an increased tyrosine kinase activity and has a 210 kDa size. The Ph1 chromosome occurs in most patients with chronic myelogenous leukemia. c-Abl can potentially regulate cell magnification and may participate in magnification regulation at multiple cellular locations, interacting with different cell components. It contains SH2 and SH3 domains (optically discrn SH domains) and withal domains involved in binding to F-acDNA, andDNA and occurs in both cytoplasmic and nuclear locations. Its DNA-binding activity appears to be cell-cycle-regulated by Cdc2- mediated phosphorylation; it binds the retinoblastoma protein betokening involution in transcriptional regulation.

Ablation

1. (in surgery) the abstraction or ravagement of tissue by a surgical procedure.
2. (in genetics) a technique used to isolate a tissue or a particular cell type during development. Genes for toxin proteins such as diphtheria A (dipA) are tissue-specifically expressed in a transgenic organism.

ABM

Aabbr. for 2-aminobenzyloxymethyl, a group used for derivatizing cellulose or paper. It is converted by diazotization into DBM.

Abortive Complex

An enzyme-substrate intricate known as a dead-end intricate or nonproductive intricate is one in which the enzyme is bound to the substrate in such a way that it renders catalysis impossible, so that no products are formed.

Abortive Transduction

This type of transduction occurs when the recipient chromosome does not integrate the donor DNA although it persists as a nonreplicating fragment that works physiologically and can be replicated in one daughter cell during each cell division.

ABO System

This system, one of the blood group systems, is of great importance to blood transfusion as human plasma contains natural antibodies to A and B blood group antigens.

Abri

In familial British dementia, a neurotoxic 34-residue polypeptide is found in amyloid fibrils originating from a mutant putative transmembrane precursor.

Abrin

Indian liquorice (*Abrus precatorius* L.), a tropical Asian vine that grows in Florida, contains a highly toxic polypeptide *65 kDa glycoprotein isolated from its seeds. Each of its two chains, A and B, is composed of an acidic 30 kDa chain and a neutral 35 kDa chain held together by disulfide bonds. Protein synthesis is inhibited by the A chain while the B chain is a carrier for binding abrin to the membrane and perhaps assisting penetration by the A chain. One well-chewed seed can be fatal. The A and B chains are derived from a common 528 amino acid 59.24 kDa precursor.

Abscisic Acid

Or (formerly) abscisin II or dormin abbr.: ABA; 5-(1-hydroxy-2,6,6-trimethyl-4-oxocyclohex-2-en-1-yl)-3-methylpenta2,4-dienoic acid; a chiral sesquiterpene. The naturally occurring form, the 2Z,4E, S isomer, additionally designated (S)-abscisic acid, is a phytohormone composed by the degradation of carotenoids. It controls abscission in flowers and fruit but probably not in leaves, and is withal implicated in geotropism, stomatal closure, bud dormancy, dormancy of seeds requiring stratifion (i.e., those that will only germinate after exposure to low temperatures), and possibly tuberization.

Absolute

1. Pristine, undiluted, e.g., absolute alcohol.
2. Not relative, e.g., absolute configuration.
3. Describing a quantification defined in fundamental units of mass, length, and time that does not depend on the characteristics of the quantifying apparatus, e.g., absolute temperature.

Absolute Alcohol

The common title for unadulterated ethanol, i.e., ethanol that has been liberated of water. It may contain little sums of benzene that have been included to help in expelling water. Substances with absolute alcohol render it unfit for human utilization and thus free of extract obligation. Industrial spirit contains 5% v/v methanol, whereas methylated spirit too contains pyridine, petroleum oil, and methyl violet color, and surgical spirit too contains castor oil, diethyl phthalate, and methyl salicylate.

Absolute Reaction Rate Theoy

Aa hypothesis hat sets out to anticipate the absolute response rate of a chemical response from the quantum mechanical depiction of the potential energy changes amid the interaction between chemical species. It is most broadly drawn upon in applying thermodynamic approaches to equilibria between reactants within the ground state and chemical species within the actuated state or dynamic state.

Absorbane

Aa degree of the capacity of a substance or preparation to assimilate electromagnetic radiation incident upon it. It breaks even with the logarithm of the proportion of the transmittance of the incident radiation, Φ_0, to the transmittance of the transmitted radiation, Φ. For a preparation, absorbance is written as the logarithm of the proportion of the light transmitted through the reference test to that of the light transmitted through the preparation, the perceptions being made utilizing indistinguishable cells. (Customarily, brilliant concentrated was measured rather than brilliant control, which is presently the acknowledgedframe.). Two amounts are characterized: (decadic) absorbance (image: A10 or A), and napierian absorbance.

Absorbed Doe

(Iin radiation material science) a degree of the energy testimony delivered by ionizing radiation in any (indicated) medium as a result of the ion-pair arrangement. The CGS unit of retained dosage is the rad; the SI determined unit is the gray.

Absorptiometer

1. A device, as often as possible a photoelectric gadget, for measuring light assimilation by solids, fluids, or gasses.
2. A device for measuring the sum of gas absorbed by a fluid.

Absorption

1. The process whereby one substance, such as a gas or fluid, is taken up by or penetrates another fluid or solid.
2. The maintenance by a material evacuated from electromagnetic radiation passing through the material.
3. The expulsion of any frame of radiation, or the diminishment of its energy, on passing through matter.
4. The method whereby a neutron or other molecule is captured by a nuclear core.
5. In cellular physiology: the up-take of liquids by living cells or tissues. In organism physiology: the totality of the forms included in causing water, the items of assimilation, and exogenous substances of moo atomic mass such as drugs, salts, vitamins, etc. to pass from the lumen of the gastrointestinal tract into the blood and lymph. I n plant physiology: the take-up of water and broken-down salts through the roots.
6. In immunology: the method of evacuating a specific counter acting agent (or antigen) from a complex by including the complementary antigen (or counter acting agent).

Absorption Band

Also known as the absorption line, is a region of darkness or absorption of radiation in the spectrum of heterochromatic radiation that has passed through an absorbing material.

Absorption Spectrum

A range created when electromagnetic radiation is absorbed by a test. The frequencies of the radiation retained are those able to energize the particles or particles of the test from their ground states to energized states. The recurrence, v, at which a specific retention line occurs depends on the energy differenceee, ΔE, between that of a specific ground state which of the comparing energized state.

Absorptivity

A degree of the capacity of a material to assimilate electromagnetic radiation. It rises to the absorptanc o a test of the material isolated by the optical path-lengthesFor approximates the retention coefficient. Utilization of the term isn't suggested.

Accelerator

1. (In chemistry) catalyst, particularly one that increments the rate of a polymerization response.
2. (In material science) a gadget or machine utilized for giving tall active energy to charged subatomic particles, e.g., electrons, protons, or alpha particles, by implies of electric and/or magnetic fields.

Accession

The number or code that uniquely recognizes a section in a specific database. Specific numbers are consigned when sections are to begin with included to a database and ought to stay inactive between overhauls, giving a dependable implies of finding them in consequent discharges. For illustration, P02699 distinguishes bovine rhodopsin within the Swiss-Prot database, and IPR000276 recognizes the rhodopsin-like G protein-coupled receptor superfamily in InterPro.

Accessory Pigment

Any of the pigments, such as the yellow, red, or purple carotenoids and the red or blue phycobiliproteins in photosynthetic cells. The carotenoids are continuously expressed, while the phycobiliproteins are present in green plants having a connetion to the Rhodophyceae, the Cyanophyceae, and the Cryptophyceae. Chlorophyll b is additionally an accessoty pigment.

Acetaminophen

Also known as paracetamol 4-acetamidophenol; N-acetyl-paminophenol; N-(4-hydroxyphenyl) acetamide; a sedate broadly utilized as for pain relieving and antipyretic. It restrains prostaglandins inside, but not exterior of the brain. It is metabolized inside the liver for the most part to glucuronide and sulfate conjugates. A little sum is oxidized to a profoundly responsive middle of the road, Nacetylbenzoquinoneimine, that's regularly detoxified by conjugation with glutathione. In case it is created in overabundance of the capacity of the liver to detoxify it, it can cause hepatic cirrhosis. It can be managed with methionine, which increases glutathione within the liver. N-Acetylcysteine is managed in cases of harming to act as a glutathione substitute. Common names are: Panadol, Tylenol.

Acetylcholinesterase

Abbr.: Throb; precise title: acetylcholine acetylhydrolase; other names: genuine cholinesterase; cholinesterase I; an esterase protein that catalyzes the hydrolysis of acetylcholine to choline and acetic acid derivation; it moreover acts on an assortment of acidic esters and catalyzes transacetylations. It is found in or joined to cellular or cellar layers of presynaptic cholinergic neurons and postsynaptic cholinoceptive cells. A change happens in cerebrospinal liquid and inside cholinergic neurons. It is hindered by certain drugs, e.g., physostigmine, and by a few organophosphates. The 3-D structure is known for parts gotten from the electric beam (fish).

Achondroplasia

The foremost common form of dwarfism, acquired as an autosomal disease. It is caused by one of two missense transformations within the fibroblast development figure receptor-3 quality (FGFR3) locus at 4p16.3, which influences the transmembrane position of the receptor, causing activation of the receptor. Homozygosity is deadly within the neonatal period. Milder forms, called hypochondroplasia, are due to any of several missense changes that influence the tyrosine kinase of this receptor. Pseudoachondroplasia is caused by over 70 changes at 19p13.1, inside the cartilage oligomeric metabolic protein (COMP). These lead to aggregation of the mutant protein inside chondrocytes.

Achromic Point

The point in time amid the activity of amylase on starch at which the test does not give a color with iodine, i.e., the response has continued to the point when the starch has all been used at slightest as distant as achröodextrins.

B

Baa Helices

Abbr. for essential amphiphilic a-helices; a-helices that contain a cluster of fundamental amino-acid buildups (His, Lys, or abbr. for essential amphiphilic a-helices; a-helices that contain a cluster of fundamental amino-acid buildups (His, Lys, or Arg) on one side and hydropathic buildups on the other. They are found, e.g., within the calmodulin-binding districts of numerous proteins.

Bacillus

A sort of huge rod-shaped Gram-positive eubacteria having a place in the family *Bacillaceae*. Its individuals are oxygen-consuming or facultatively anaerobic, spore-bearing organisms. B. *subtilis* has gotten to be broadly set up as a vehicle for hereditary designing and numerous cloning vectors are accessible. *B. thuringiensis* synthesizes a poison that's dynamic against insects; diverse strains of the bacterium create diverse shapes of the poison particular for distinctive insect species.

Figure 2. Photomicrograph Of *Bacillus anthracis* Bacteria.

Source : Image By Pixnio

Bacitracin

A cyclic anti-microbial polypeptide complex formed by *Bacillus subtilis* and *B. licheniformis*. Commercial bacitracin could be a blend of at slightest nine bacitracins, for the most part, bacitracin A. Bacitracin A contains, in addition to a few L amino acids, a number of D amino acids. Bacitracin hinders bacterial cell wall formation by blocking the dephosphorylation of undecaprenyl diphosphate (undecaprenyl pyrophosphate) to the monophosphate frame; it essentially represses the dephosphorylation of dolichyl diphosphate subsequently blocking the arrangement of the center oligosaccharides of glycoproteins.

Back Titration

A roundabout titration strategy in which a measured overabundance of the reagent is included and the sum remaining after response with the analyte is titrated back to antendpoint, the extent used within the response being gotten by differences.

Bactenecin

Any of a few profoundly cationic polypeptides, initially separated from bovine neutrophil granules, but moreover found in sheep, that apply, *in vitro*, a strong antimicrobial action, conceivably owing to restraint of the respiratory chain.

Bacteria

One of three superkingdoms of cellular life forms, the others being *Archaea* and Eukarya. Microscopic organisms are unicellular and anucleate i.e. prokaryotes. They document huge differences of shapes, major divisions counting the *Cyanobacteria*, *Proteobacteria* (which incorporates Gram-negative microbes), and Gram-positive bacteria.

Bacteriochlorophyll

Any of the chlorophylls of photosynthetic microscopic organisms. They contrast fundamentally from the chlorophylls of higher plants. Bacteriochlorophylls a to g are known. Bacteriochlorophylls a and b are the leading known, being the photosynthetic colors of purple microbes. Their purple color comes about from the truth that they are reduced in both rings B and D, and hence may be respected as tetrahydropyrroles. Tetrapyrrole carbon particles and carbon rings are numbered agreeing to the Fischer and IUPAC systems.

Bacterioferritin

abbr: BFR; other names: cytochrome b1 or cytochrome b557; an iron-storage protein comprising 24 indistinguishable subunits that pack to create a symmetrical, near-spherical shell, encompassing an ~8 nm central depth. imilitudenear similitude between BFR and the iron-storage ferritin found in eukaryotes and microbes. BFRs store expansive amounts of press within their empty insides (13–20% w/w of iron), within a hydrated ferric oxide mineral containing variable levels of a phosphate anion.

Bacteriorhodopsin

A retinal-containing protein, organismformed by *Halobacterium halobium* and other halophilic archaebacteria, and embedded into patches of purple layer within the cell surface. The purple films serve as light-operated proton pumps to translocate protons from the interior to the exterior of the cells.

The mature protein may be a 7TM protein of known 3-D structure; until the structure of bovine rhodopsin was decoded, this given a show for the engineering of opsin and related G protein-coupled receptors in eukaryotic cells.

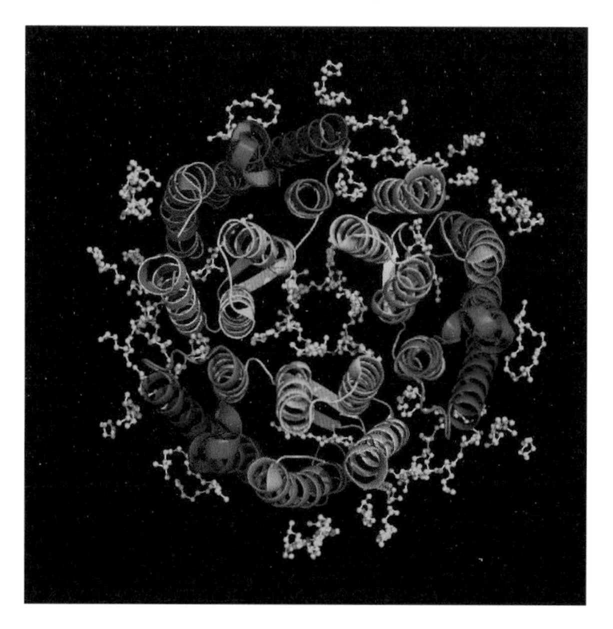

Figure 3: Ribbon Model Of Bacteriorhodopsin.

Source : Image By Wikimedia Commons

Bacterium

(pl. microscopic organisms) any of a tremendous and ubiquitous group of prokaryotic microorganisms that exist as single cells or in clusters or groups of single cells. Most specialists presently put them within the exclusively prokaryotic kingdom Monera, at the side the *cyanobacter*een greengreen, green alga). Th part ofpart of microbes have an inflexiall;thosewall; those missing this include are named Archaea. Micoscopic oorganisms, Archaea, and Eukarya constitute the three essential Kingdomcterial adj.

Baculovirus

A group of DNA containg pathogennns that are known to increase as it were in organism invertebrates and are presently classiied within the family

Baculoviridae. Their genome comprises of double-stranded circular DNA of 58–100 MDa. T extend they have potential as pest-control operators. Baculovirus vectors are profitable in expressing certain organism proteins.

Figure 4. Baculovirus on.

Source :Source: Image By Flickr

Balanced growth

A sort of development s such thatiee interim (insidhexponntial stage). Balanced growth is a vital attribute of all livingsamThe Adjusted development for an individual uresrequires that each cell after division is a correct copy of the one e past cyclecycles.

Barbiturate

1. The tautomeric anion derived from barbituric acid (malonylurea; 2,4,6(1H,3H,5H)-pyrimidinetrione).
2. Any coalescence of free barbituric acid and its anion.
3. Any salt of barbituric acid.

4. Any of many pharmacologically active derivatives of barbituric acid, including barbitone, pentobarbitone, phenobarbitone, and thiopentone. Barbiturates are potent CNS depressants that potentiate the actions of GABA by binding to the GABAA receptor. Pentobarbitone and thiopentone are acclimated to induce anesthesia.

Barcroft Apparatus

The Barcroft respirometer, now rarely used for studying gas exchange in cells, tissues slices, or tissue homogenates, makes use of differential respiration to study gas exchange. It is a closed system comprising two flasks of equal volume connected by a U-shaped manometer (the Barcroft manometer). As both flasks condense liquid and gases, the reaction flask contains cells or tissues; the emolument flask, on the other hand, is free from cells or tissues and adjusts to compensate for changes in barometric pressure and temperature during the experiment. The respirometers are arranged so that they may be shaken with the flasks in a constant-temperature bath. The reading of the apparatus is the distinction between the calibers of the manometer fluid in the two sides of the manometer.

Basal Lamina

A thin sheet of proteoglycans and glycoproteins, especially laminin, secreted by cells as an extracellular matrix, composing a region between the cells and adjacent connective tissue. The basal lamina has consequential functions in the organization of tissues including an influence on cell polarity, cell differentiation, and cell migration. The basal lamina is subdivided into the lamina lucida (an electron-lucent region) immediately adjacent to the cell layer, and the lamina densa (an electron-dense region) external to it. Frequently there is a further outermost layer, the lamina reticularis, containing collagen fibrils.

Base Analog

(Esp. US) base analog any unnatural purine or pyrimidine base that can be incorporated in vivo into DNA structure. This genetic alteraction can induce different properties by altered base-pairing during incorporation or in subsequent DNA replication. Thus 5- bromouracil, an analog of thymine, pairs with guanine thereby causing transitions of A–T → G–C. Similarly, 2-aminopurine, an analog of adenine, pairs with cytosine thereby causing transitions of A–T → G–C. Some base analogs have been utilized as anticancer and antivirus agents.

Basement Membrane

A thin layer of dense material found in many animal tissues interposed between the cells and adjacent connective tissue. It consists of the basal lamina, composed mainly of a tightly cross-linked meshwork of type IV collagen molecules interwoven with a laminin network, to which it is bound by entactin molecules. The collagen molecules are glycosylated, composed of largely triple-helical amalgamations of a chains, which may be of six types: a-1(IV) to a-6(IV). There is an associated layer of reticulin fibers present in various tissues in the body, for example,, capillary linings, kidney tubules, lung alveoli, and renal glomeruli. basement membrane has an ancillary function in some tissues and may withal act as a passive selective filter for substances diffusing in or out of the cells, e.g., in renal glomeruli, it retains protein molecules. It gives a vigorous periodic acid–Schiff reaction.

Basophil

Also known as basophil(ic) leukocyte, is a polymorphonuclear phagocytic leucocyte of the myeloid series that is distinguished by the presence of coarse cytoplasmic granules that stain with rudimental dyes. The granules are believed to contain histamine, heparin, and other vasoactive amines. Basophils are proximately cognate to mast cells.

Bence-Jones Protein

A group of myeloma proteins, occurring in the urine of patients with multiple myelomatosis and cognate conditions, that precipitate on heating the urine to 60 °C, but redissolve at higher temperatures. They consist of light chains of immunoglobulins synthesized by the myeloma cells.

Benedict's Solution

A reagent used to test for the presence of glucose and other reducing sugars. When heated with reducing sugars it is reduced and a red precipitate of cuprous oxide is formed. There are discrete formulations for qualitative and quantitative analysis.

Benzoic Acid

Also known as benzene monocarboxylic acid a white crystalline solid found naturally in plants and additionally utilized as a food preservative. In plants, it is derived metabolically from phenylalanine. It accumulates (with salicylic acid), free and as a glycoside, in the immediate vicinity of infection sites. In animals and humans, it is excreted in urine after conjugation to benzoylglycine (or hippuric acid) in the liver.

Biacore

he proprietary name of an optical biosensor instrument that measures authentic-time interactions of macromolecules utilizing surface plasmon resonance. Ligands are immobilized on a dextran-coated gold surface. The resulting 'sensorgrams' can be adjusted to derive kinetic parameters of the interactions.

Bibliographic Database

A database that houses bibliographic information, primarily in the form of citations, controlled lexicon terms, index codes and abstracts. Examples include MEDLINE, which covers the ecumenical biomedical literature; EMBASE Excerpta Medica, which covers the ecumenical biomedical and pharmaceutical literature; and PASCAL, which covers the major international literature on science, technology, and medicine.

Bigdye Terminator

A proprietary name for reagents utilized in DNA sequencing in which the dideoxynucleoside triphosphates are each labeled with a different fluorescent dye to facilitate their identification when the products of sequencing reactions are dissevered by gel or by capillary electrophoresis.

Bikunin

A plasma glycoprotein found both in the free state and complexed with the heavy chains of the inter-a-inhibitor family, composing the light chain. It is a serine protease inhibitor, having a tandem arrangement of Kunitz domains; it may participate in the control of events such as endothelial cell division, and oocyte cumulus expansion and stabilization. It is synthesized from an α-1 -microglobulin/ bikunin precursor.

Bilene

The semisystematic name for either of two tetrapyrroles in which the carbon bridges contain one more double bond than bilane; i.e. one of the three bridges in the molecule is a methine group.

Bioautography

A method for detecting, in an involute coalescence, mixes of substances that are able to act as essential magnification factors for a test organism(s). The components of the mixture are first separated by chromatography. The resulting chromatogram is then placed in contact with a culture of the test organism in a solid medium without an essential nutrient. Hence, the organism will only grow where the component is present in the chromatogram, thereby establishing its presence in the pure form.

Biochemistry

For biological chemistry the branch of science dealing with the chemical compounds, reactions, and other processes that occur in living organisms. Lehninger expressed the challenge to the biochemist as follows: 'Living things are composed of desolate molecules. When these molecules are isolated and examined individually, they conform to all the physical and chemical laws that display the behavior of inanimate matter. Yet living organisms possess extraordinary attributes not shown by amassments of inanimate molecules.' In this regard Horowitz has proposed a set of criteria for living systems: 'Life possesses the properties of replication, catalysis and mutability.' Biochemists are, consequently, concerned with the manner in which living organisms exhibit these properties.

Bioengineering

1. Engineering relating to the operation on an industrial scale of biochemical processes, especially fermentation. This is conventionally now termed biochemical engineering.
2. The application of the physical sciences and engineering to the study of the functioning of the human body and to the treatment and redressment of medical conditions.

Bioinformatics Institute

A research and accommodation centre that engenders and manages biological databases (e.g. of nucleic acid and protein sequences, protein families, macromolecular structures), and develops software (e.g. for analyzing protein sequences and structures, genome sequences, microarray data) for free use by the community. The European Bioinformatics Institute (EBI) and the National Center for Biotechnology Information (NCBI) are the primary bioinformatics research centers for Europe and the USA respectively.

Biophysics

The application of physical techniques and physical methods of analysis to biological quandaries. Traditionally, the discipline has fixated on two main areas: first, the transmission of nerve signals and the maintenance of electrical potentials across membranes; and second, molecule crystallography and enzyme structure and mechanisms. For the latter X-ray crystallography has been supplemented by a number of physical techniques, including nuclear magnetic resonance spectrometry, mass spectrometry, fluorescence-depolarization quantifications, and circular-dichroism studies.

Biotinidase

EC 3.5.1.12; Enzyme involved in the removal of biotin from biocytin or biocytin peptides but not biotin holocarboxylases. It is present in many human tissues. Serum biotinidase is a glycoprotein with about nine isoforms (67–76 kDa) and is derived from the liver. Many mutations in a locus at 3p25, which encodes a mature protein of 502 residues, result in biotinidase deficiency. This is an autosomal recessive disorder of highly variable clinical expression that commonly includes seizures, hypotonia, developmental delay, ketolactic acidosis, and organic aciduria. Serum biotinidase activity is below 10 to 30% of normal.

Bleomycin

Any of a gather of related glycopeptide anti-microbials separated from Streptomyces verticillus. Bleomycin acts to halt the cell cycle within the G2 stage (see cell-division cycle). It is utilized to actuate synchrony in cell systems and as an antineoplastic specialist, particularly in lymphomas.

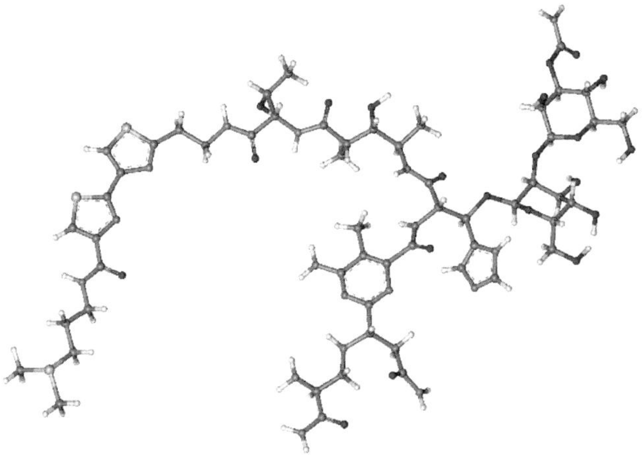

Figure 5. Ball-And-Stick Model Of The Bleomycin Molecule.

Source : Image By Wikimedia Commons

Blue Shift

Any move of the crests of retention or emanation of a range of electromagnetic radiation to shorter wavelengths, counting from the obvious to the ultraviolet region.

Bohr Effect

The change in oxygen affinity of hemoglobin with pH (the oxygen binding increments with expanding pH). It is one of the impacts emerging from oxygen-linked acid groups in hemoglobin and comparative oxygen-carrying proteins conjointly includes comparable impacts including other acid-linked capacities, e.g. oxidation damage of protons causing destruction. The impact was found by Christian Bohr et al. taking after observations that changes within the partial pressure of CO_2 influence the oxygen balance of the blood.

Bone Marrow

The soft tissue contained within the internal cavities of bones. Red blood cells are active hemopoietic cells found within the marrow (or myeloid tissue), which is found within long bones, ribs, and vertebrae. The blood cells are accompanied by megakaryocytes, reticulum cells, macrophages, and plasma cells. In adult animals the marrow of many bones, concretely limb bones, becomes filled with fatty tissue kenned as yellow marrow. In adult mammals, B lymphocytes develop and differentiate in the bone marrow.

Bottom Yeast

The popular name for any of many strains of brewer's yeast that effect fermentation at a comparatively low temperature and incline to sediment to the bottom of the fermentation vessel. It is utilized for the manufacture of light potations, e.g., lager.

Bouvardin

A cyclic hexapeptide isolated from the plant *Bouvarda ternifolia*, utilized as a drug against dysentery. It has anti-tumor properties and inhibits eukaryotic protein synthesis.

Branched-Chain Amino Acid

Abbr.: BCAA; any of the neutral aliphatic essential amino acids L-leucine, L-isoleucine, and L-valine. They are incorporated into proteins or degraded oxidatively in mitochondria, especially in skeletal muscle, liver, and encephalon, and are ketogenic. BCAA infusions counteract the catabolic state of astringent trauma and sepsis. Maple-syrup urine disease results from their defective oxidative decarboxylation.

Branch Migration

A proposed model to expound the occurrence of branched DNA structures, as optically discerned by electron microscopy, and their conversion into linear duplex DNA. The branched structures may form by sodality over components of their lengths of three or more polynucleotide strands. The migration of the branch point is postulated to occur by displacement of a strand from its fellow in one branch by a strand in another. Such a process may occur in transcription or in recombination.

Brevetoxin

Ptychodicus brevis (*Gymnodinium breve*) is a red tide organism that produces highly toxic polyether compounds that are lipid soluble. They cause the death of fish and are responsible for shellfish poisoning in humans. They are activators of voltage-dependent Na^+ channels.

C

Cachectic Factor

It is a 24 kDa proteoglycan that is isolated from the murine adenocarcinoma MAC16 that has been implicated in the production of cachexia. The peptide has the N-terminal sequence YDPEAASAPGSGDPSHEA, and has N- and O-glycans. Intravenous injection of the proteoglycan induces rapid weight loss.

Cachexia

It is defined as a condition that is caused by chronic disease, such as cancer. Further, it is characterized by wasting, emaciation, feebleness, and inanition. Eventually, it led to the name 'cachectin' for the protein which is now known as tumor necrosis factor.

Cadaverine

It is a substance that is formed by microorganisms that decay in meat and fish by the process of decarboxylation of lysine. In addition, it also occurs as an intermediate in the process of biosynthesis, via lysine, of some quinolizidine alkaloids (e.g., lupinine) in plants.

CAF1

Abbreviated as- chromatin assembly factor 1. It is a complex of p150 and p60 subunits and are present in Drosophila and human cells. These are responsible for chaperoning histones H3 and H4 to DNA during the process of replication. The p60 gene maps to a locus on chromosome 21 that is strongly linked with the major features of Down syndrome.

Caged ATP

It is a kind of protected ATP analog, e.g. adenosine-5′- triphospho-1-(2-nitrophenyl)ethanol (or the methanol-containing equivalent). It releases ATP in good yield when photolyed by a short pulse of light of 360 nm wavelength. In a similar manner, guanosine-5′- triphospho-1-(2-nitrophenyl)ethanol is used as caged GTP. These compounds can further be introduced into cells prior to the process of photolysis. Caged ATP has been used in order to study muscle contraction on the millisecond timescale.

Calcidiol

Calcidiol is the recommended trivial name for calcifediol, 25-hydroxycholecalciferol. It is formed in the liver from cholecalciferol. It is the major store of the vitamin in the body that is present mainly in plasma. It is the precursor of the hormonal form of the vitamin, calcitriol.

Calcitonin Receptor

Calcitonin receptor is a membrane protein that binds calcitonin and mediates its effects. They are seven-transmembrane-helix receptors, generally coupled to G-proteins. In numerous systems, they can activate both adenylate cyclase as well as the phosphatidylinositol cycle.

Calcium Phosphate-Mediated Transfection

Calcium phosphate-mediated transfection is a method for introducing exogenous DNA like a plasmid into mammalian or other eukaryotic cells. Cells in culture are exposed to a calcium phosphate coprecipitate of DNA and a part of them take up the complex by endocytosis and express the transfected genes. Transfection efficiency varies greatly with different cell lines.

Calcium-Sensing Receptor

Calcium-sensing receptor is a G-protein-coupled receptor that is expressed in parathyroid glands and renal tubules. It contains 1078 amino acids, encoded by a gene at 3q13.3-q21. Loss-of-function mutations are associated with familial hypercalcemic hypocalciuria, and gain-of-function mutations are associated with familial hypocalcemic hypercalciuria.

Calculus

(pl. calculi) A concretion of material that forms inside the body. It generally resembles a small pebble, therefore is called a 'stone'. Calculi are most common in the gall bladder or kidney, and are composed variously of organic or inorganic salts, frequently of calcium; cholesterol calculi are gallstones of pure cholesterol.

Clcyclin

Oother names: prolactin receptor associated protein, growth factor-inducible protein 2A9, S100 calcium-binding protein A6; a small protein which copurifies with prolactin receptor. Calcylin is induced in fibroblasts by growth factors and is overexpressed in acute myeloid leukemia.

Caldesmon Kinase

EC 2.7.1.120; the name given to a reaction now known to result from the autophosphorylation of caldesmon (by ATP) in the presence of calcium. This autophosphorylation abolishes the ability of caldesmon to bind actin.

Calnexin

Calnexin is a calcium-binding protein of the endoplasmic reticulum. It seems to play a part in the processing of endoplasmic reticulum proteins, in monitoring assembly, and in retaining unassembled or incorrectly folded proteins. Calnexin has a single transmembrane helical region.

Calpastatin

Calpastatin is a protein, found in liver and erythrocytes of several mammals, which is a specific calpain inhibitor. Four characteristic domains are involved in the inhibitory action.

Calponin

Calponin is a thin filament-associated protein associated in the regulation and modulation of smooth muscle contraction. Calponin is found as an actin-, calmodulin-, and tropomyosin-binding protein present in many vertebruscles, andmuscles and is related to troponin T in immunological and biochemical characteristics.

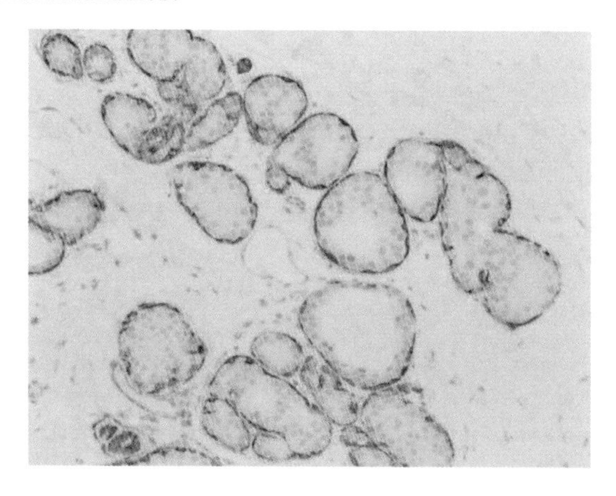

Figure 6. Immunohistochemistry With Calponin In Ductal Carcinoma In Situ.

Source : Image By Wikimedia Commons

Campomelic Dysplasia

It is also known as autosomal sex reversal, which is a rare autosomal dominant chondrodysplasia. It is usually caused by mutations in SOX9 and fatal in early infancy. It is associated with feminization in most affected genotypic male patients.

Canavan Disease

It can be defined as an autosomal recessive degenerative disease. This is mainly caused due to a defect in the gene for aspartoacyclase. Onset is in early infancy, entailing blindness, psychomotor regression, enlarged head, hypotonia, optic atrophy, spasticity, and increased urinary excretion of N-acetylaspartate.

Candida

A genus of (in most of the cases) dimorphic yeasts containing species of medical (*C. albicans*) and industrial importance: *C. cylindrica*. It is a source of lipases. In addition, it is a rare example of a eukaryote with a non-standard genetic code for nuclear/cytoplasmic gene expression.

Capecitabine Or Xeloda

Chemically it is 5′-deoxy-5-fluoro-N-[(pentyloxy)carbonyl]-cytidine, that is the first tumor-activated drug. It principally acts by inducing the tumor to produce more thymidine phosphorylase, which in turn causes the tumor to make more cytotoxic 5-fluorouracil, to which the drug is converted.

Capillarity

Capillarity is the phenomenon, resulting from surface tension, in which fluids ascend capillary tubes and makes them form a concave or convex meniscus at their surface where they contact a solid.

Capsular Antigen

Capsular antigen or capsular polysaccharide or capsular substance any of the antigens, generally polysaccharide in nature, which are carried on the surface of bacterial capsules. The term K antigen is utilized for those that mask somatic (O) antigens.

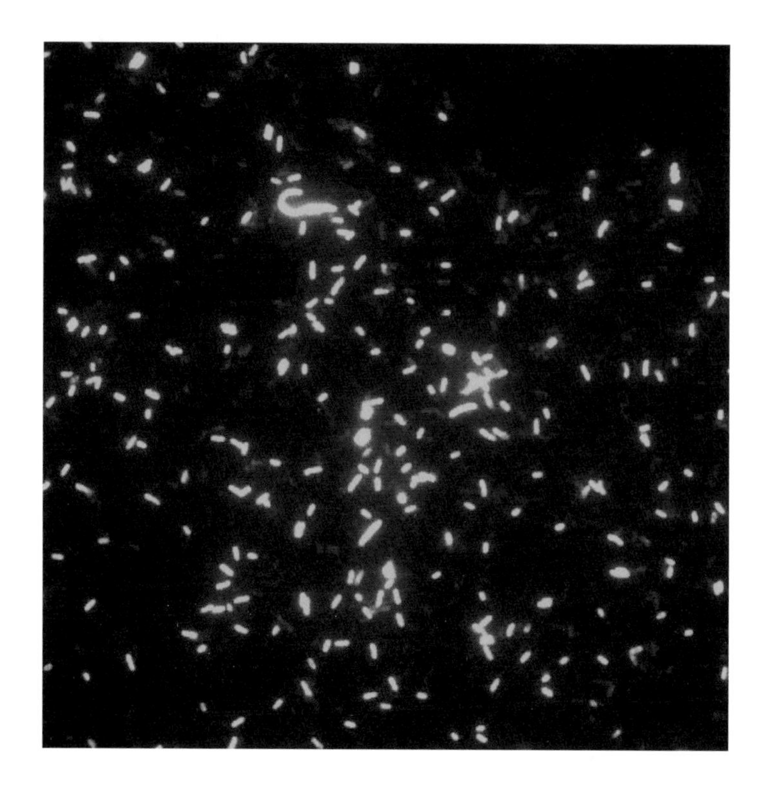

Figure 7. Plague *Yersinia pestis*, Capsular Antigen.

Sources : Image By Pixnio

Carbon Cycle

Carbon cycle is the aggregate of chemical and biochemical processes by which carbon is cycled between carbon dioxide in the environment (and sea) and organic compounds in organisms and their remaining part. Atmospheric carbon dioxide is fixed during photosynthesis by green plants and other photoautotrophic organisms, utilizing sun-based energy to split water molecules liberate O_2. The carbon dioxide is in this manner is changed into glucose and other organic compounds, that are utilized as metabolic fuel and building blocks by most of the organisms. Aerobic organisms, specially, are capable completely to oxidize organic compounds during respiration to release carbon dioxide into the atmosphere. The carbon kept in the remaining part of dead organisms is liberated by the respiration of decomposers or, particularly that in fossil fuels, by combustion. Thus, photosynthetic organisms and aerobic heterotrophic organisms exist in a condition of s. GStill, green plants and other aerobic autotrophs also liberate carbon dioxide during the respiration process.

Carbon Dating

Carbon dating or radiocarbon dating is a procedure for ingdeciding the age of specimens of biological origin, for example: wood. It depends oingdeciding how much carbon-14 remains in the specimen, the known half-life of carbon-14, and the assumption that the abundance of this radionuclide in the atmosphere has stayed constant as its incorporation into the material from atmospheric CO_2.

Celiac Disease

Or (Brit.) coeliac disease or gluten enteropathy or nontropical sprue is a situation produced in genetically susceptible individuals by damage to the mucosa of the upper small intestine triggered b the alphaa-gliadin component of gluten in wheat and e Alpha-a gGliadin is rich in glutamine residues, and the situation is set apart by the presence of circling antibodies to tissue transglutaminase. This enzyme is released by mechanical or inflammatory stress, generally, this is in an inactive form in submucosal cells of the upper small intestine. The enzyme deamidates specific glutamine residues in peptides derived from a-gliadin hence producing peptides that activate the T cells of the submucosa. The disease is described by diarrhea and malabsorption, however, can be relieved by the elimination of gluten from the diet.

Cell-Free System

A cell-free system is any experimental system made out of subcellular fractions and/or cell-free extracts. Cell-free systems for amino-acid incorporation (protein synthesis) normally comprise ribosomes, natural or synthetic mRNA, tRNAs, enzymes, amino acids, an ATP generating system, GTP, buffer, certain inorganic salts, and few organic compounds.

Cell Fusion

Cell fusion is the creation of a single hybrid cell containing the nuclei and cytoplasms from various cells. It very well might be induced by treatment of a mixed cell population with certain fusogens (for example: killed Sendai virus or polyethylene glycol). It is a significant step in the strategy of forming hybridoma cells for the creation of monoclonal antibodies.

Cell-Mediated Immunity

Cell-mediated immunity is a specific immunity that relies upon the presence of T lymphocytes. It is liable for, e.g., allograft rejection, delayed hypersensitivity, and tuberculin test reactions, and is significant in the organism's defense against viral and a few bacterial infections.

Cellodextrin

Cellodextrin is a glucan chain linked to sitosterol. It is mainly formed by the transglucosylase components of cellulose synthase in plant cell plasma membranes, by the transfer of the glucosyl moiety of UPD-glucose onto the sitosterol beta-glucoside primer. The glucan is then released by Korrigan cellulase to form the microfibrils ofse and alsoand to regenerate the sitosterol b-glucoside primer.

Cellulose Synthase

It is an enzyme complex that is present in plant cell plasma membranes. The key function is to synthesize cellodextrin by transfer of glucose from UDP-glucose to a sitosterol b-glucoside primer. First of all, the primer is synthesized by a component of the complex by the transfer of glucose from UDP-glucose to sitosterol. The elongating transglucosylation reactions are further catalyzed by different subunits that surround the primer synthase. Cleavage of the cellodextrin from the sitosterol beta-glucoside primer is achieved by Korrigan cellulase, itself a component of the complex.

Cell Wall

A semi-rigid or rigid envelope present outside of the cell membrane of plant, fungal, and most of the prokaryotic cells. This helps in maintaining the shape of the cell as well as protect them from osmotic lysis. In the case of prokaryotes, it is present inside the capsule and slime layer. It mainly comprises peptidoglycan. It can be removed by using different kinds of techniques with retentionof its 3-Dd form. In addition, in the case of fungi, the cell wall is largely made up of polysaccharides, on the other hand, in plants, it is made up of cellulose and lignin.

Centrifugal Elutriation

Centrifugal elutriation is a method where cells or other particles are isolated by elutriation in an extraordinarily built centrifuge rotor; the increased gravitational field resulting from centrifugation speeds up the division of the particles.

Centromere-Binding Proteins

These are proteins from, e.g., the yeast kinetochore that are capable of binding centromere DNA; examples from yeast: centromere-binding protein 1, a helix-turn-helix protein that binds to a conserved DNA sequence termed CDE-I. Additionally, it is involved in chromosome segregation; centromere-binding protein 5, which is involved in mitotic chromosome segregation. It is well-known that these proteins are essential for cell growth.

Cerebrospinal Fluid

Abbreviated as CSF, a clear fluid that contains little protein and few cells fills the subarachnoid space and ventricles of the brain, along with the central canal of the spinal cord. Around 80% of the protein is derived from plasma, while on the other hand, the rest is brain-specific and comprises myelin basic protein, glial fibrillary acid protein, creatine kinase (BB isozyme), and neuronal enolase.

Chaikoff Homogenizer

Chaikoff homogenizer is a hydraulically functioned tissue or cell homogenizer in which the sample is constrained through an annulus by a piston. The diameter of the piston relative to that of the annulus is picked to suit the size of the part that it is preferred to segregate with minimal damage.

Chain Reaction

Chain reaction is a chemical reaction where at least one reactive intermediate is continuously regenerated, frequently through a repetitive cycle of elementary 'propagation' steps.

Channels-Ratio

Channels-ratio is a technique of quench correcting in liquid scintillation counting where two channels are utilized to measure the average energies of beta particles both before as well as after quenching.

Chaotropic Dissociation Assay

Chaotropic dissociation assay is a method utilized to measure the heterogeneity of antibody affinities in, for example, serum. Immune complexes are dissociated in buffers of varying strengths and pH or in chaotropic agents.

Chaperone

Any of a functional class of unrelated families of proteins that help the correct non-covalent assembly of other polypeptide containing structures *in vivo*, however, are not parts of these assembled structures while they are performing their normal biological functions.

Chemical Coupling Hypothesis

According to the hypothesis, the coupling of ATP synthesis to oxidation in oxidative phosphorylation is mainly because of the formation of common 'high-energy' intermediates at the time of electron transport that are subsequently used to phosphorylate ADP to ATP. It has been superseded by the chemiosmotic coupling hypothesis.

Chemical Equivalent

Chemical equivalent is the combining proportion of a substance, by mass, relative to a hydrogen standard. It is the number of grams of that element which will combine with or replace 1 g of hydrogen.

Chemical Score

The chemical score is a measure of the nutritional value of a protein. The limiting essential amino acid in the test protein is stated as the percentage of the quantity of the same amino acid present in egg albumin, the nutritionally perfect protein. Chemical score and biological value are numerically equal.

Chemical Shift Anisotropy

Chemical shift anisotropy is a method based on nuclear magnetic resonance in this method the chemical shift changes with the orientation of the sample in the magnetic field, utilized in molecular dynamics ons and alsoand in some other fields.

Chlorinate

1. Treat or reaction (a substance) with dichlorine.
2. Introduce one or more chloro groups into an organic compound, either by addition or by substitution.
3. DDisinfection (mainly water) with dichlorine.

Chlorophyllide

A chlorophyll lacking the terpenoid side chain, generally phytol. The penultimate intermediate in chlorophyll biosynthesis. The formation of chlorophyllide is by the light-catalyzed reduction of protochlorophyllide attached to a protein, protochlorophyllide holochrome.

Cladistics

Cladistics is phylogenetic studies aimed at establishing evolutionary relationships between various species or other groupings based on the identification of natural clades.

Clearing Factor

Clearing factor is another name for lipoprotein lipase; an enzyme found bound to heparan sulfate on the capillary endothelium of several tissues, especially heart, lung, skeletal muscle, and adipose tissue, so titled because it clears the opalescence from lipemic plasma. It is released into plasma by intravenous injection of heparin.

Clostridium

Clostridium is a genus of spore-forming rod-shaped bacteria of the family *Bacillaceae*. Its members are chemoorganotrophic, obligately anaerobic, and naturally Gram-positive. They are widespread in mud, soil, and in the intestinal tract of animals. Some of the species are pathogenic if they gain access to the tissues where they can produce a range of toxins.

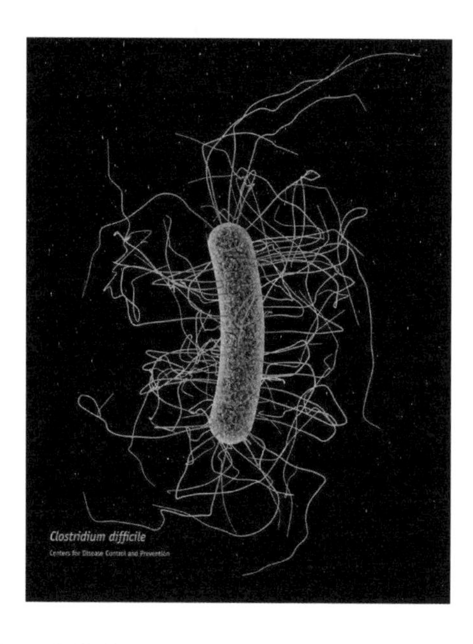

Figure 8. Clostridium difficile.

Source : Image By Wikimedia Commons

D

Damascenine

This is 3-methoxy-2(methylamino) benzoic acid methyl ester, which is a proto-alkaloid from the seeds of *Nigella damascena* (love-in-a mist). Also, it is the odoriferous principle of the oil of nigella.

Dansylate

Dansylate can be referred to as the process to derivatize with a dansyl group which is done by reacting with dansyl chloride, 5(dimethylamino)naphth-1-ylsulfonyl chloride. Such a method is used in order to acylate free amino groups in protein end - group analysis. The dansyl amino acids may be detected, which are isolated after hydrolysis of the protein and are said to be highly fluorescent and amounts as small as 1 nmol.

Danysz's Effect

Danysz's effect is also known as Danysz's phenomenon and that is the variation in the toxicity of toxin–antitoxin mixtures which depend on whether toxin, which is added to equivalence, is added in one lot (non - toxic mixture produced) or in smaller lots otimea period (toxic mixture produced). When it is added in small lots, the toxin tends to react with more than one equivalent of anti - toxin (anti-body) so that insufficient antibody is left which further neutralizes all the subsequent lotsoxintoxins.

Dapsone

This is 4,4′-diaminodiphefonesulfones; bis(4-aminophenyl) sulfone. Dapson is said to be an antibacterial and antiprotozoal substance that is used especially against *Mycobacterium leprae. Mycobacterium leprae* is the causative agent of leprosy. Dapson is also used as an adjunct in the treatment of malaria. Moreover, it acts as a competitive antagonist of p-aminobenzoic acid and an inhibitor of folate synthesis.

Daunomycin

It is also referred to as daunorubicin which is an anthracycline antibiotic and antineoplastic that is isolated from fermentation broths of *Streptomyces peucetius* and are widely used in the treatment of cancer. Furthermore, it is a glycoside that is formed from the tetracyclic aglycon daunomycinone ($C_{21}H_{18}O_8$) and the aminohexosedaunosamine. Also, it inhibits DNA replication which is done by intercalation into duplex DNA and inhibits RNA transcription.

Deaminase

Deaminase is referred to any of the enzymes that catalyze the non - oxidative removal of amino groups, along with the production of ammonia. Usually, they are aminohydrolase enzymes (sub-subclass EC 3.5.4), which hydrolytically deaminate amino-substituted cyclic amidines including various pterins, purines, or pyrimidines, either free or combined. Some kinds of hydro-lyase (sub-subclass EC 4.2.1) and hydrolytic or nonhydrolytic ammonia-lyase (sub-subclass EC 4.3.1) enzymes are commonly known as deaminases.

De-Ashing

De - ashing is a procedure in which interfering inorganic (that forms ash) components are removed from samples of carbohydrates that are being analyzed by liquid chromatography. This is done by passing them first through anion - and cation - exchange resins.

Decapentaplegic

Decapentaplegic is a secreted protein of the transforming growth actor Bb super – family, which tends to trigger different morphogenetic processes in *Drosophila*. Further, it participates in an intercellular signaling pathway which activates a particular transcription factor that is required for normal embryonal development. A homolog of decapentaplegic occurs in mammals.

Decarboxylation

This is referred to as the act or process of removing the carboxyl group from a carboxylic acid as carbon dioxide. The reaction, that takes place, can be enzyme - catalyzed by a decarboxylase or, in some of the cases, particularly with 2-oxo acids, it can be spontaneous.

Decontamination

Decontamination can be referred to as the act or process of removing or neutralizing any toxic or potentially toxic materials, or of decreasing their concentrations to non - hazardous levels. The term 'decontamination' is used esregard toregarding microbes, carcinogens, toxic chemicals, and radioactive substances.

Defensin

It is any of a family of small cationic peptides that contains six cysteines in disulfide linkage. These have broad - spectrum antibiotic action, and moreover, contribute to host defense against microorganisms. They are present in a large number in phagocytes, in the small intestinal mucosa of humans and some other mammals, and in the hemolymph of insects. Defensins further adopt multimeric pore-forming complexes in membranes. This in turn renders the membrane permeable. They are differentiated by having a predominantly beta-sheet structure.

Degradation

As mentioned in Chemistry, degradation can be referred to as:

1. The gradual stepwise as well as deliberate conversion of a molecule into smaller chemical entities, to elucidate its chemical structure.

2. Any undesired breakdown of a molecule or material along with impairment or even loss of some of its characteristic properties.
3. Depolymerization or decomposition.
4. According to biochemistry, it can also be known as catabolism.

Degradosome

Degradosome is a multi-enzyme complex that is involved in the degradation of mRNA in *E. coli*. It contains RNase E (that is the microbe's main endoribonuclease), polynucleotide phosphorylase, the glycolytic enzyme enolase, and a DEAD-box RNA helicase. Considering certain conditions, it can also contain poly(A) polymerase, polyphosphate kinase, and another DEAD-box RNA helicase.

Dehydratase

This is formerly known as the process to dehydrase any hydro-lyase enzyme, of sub-subclass EC 4.2.1, which in turn catalyzes the (reversible) breakage of a carbon–oxygen bond. This can result in the formation of an unsaturated product and the elimination of water; foleexample, citrate dehydratase, EC 4.2.1.4. When it was named after the alternative, that is the unsaturated, substrate, the enzyme is referred to as e;hydratase, such as aconitate hydratase, EC 4.2.1.3.

Denaturation

There are many definitions to explain denaturation.

1. Denaturation is referred to as (of a protein) a process in which the three - dimensional shape of a molecule is changed/modified from its native state without rupturing the peptide bonds. Sometimes, it is considered so as to include disulfide bond rupture or chemical modification of some specified groups in the protein. Denaturation is frequently irreversible and, it is accompanied by loss of solubility (specifically at the isoelectric point) and of biological activity.
2. Considering a nucleic acid, denaturation can also be known as a process whereby a molecule is converted from a firm, two-stranded, helical structure to a flexible, single-stranded structure.

Dendritic Cell

This is a type of cell that is derived from bone marrow and is characterized by branching projections and present in lymphoid tissue, skin, and squamous epithelium. It is the major cell that presents antigen in the human body. This is done by engulfing protein antigens, cleaving them, and displaying the fragments on the cell surface with the help of the major histocompatibility complex (MHC) molecules.

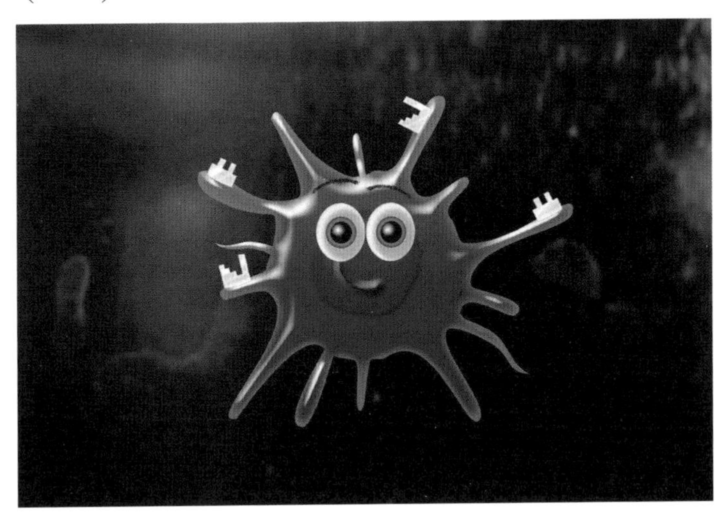

Figure 9. Supercytes - Dendritic Cell.

Source : Image By Flickr

DeoxycholicAcid

This is 3a,12a-dihydroxy-5b-cholan-24-oic acid. Deoxycholic acid is one of the bile acids which occurs as its conjugate with glycine or taurine in the bile of a number of mammals, which includes dogs, goats, humans, oxen, sheep, and rabbits. The sodium salt (that is sodium deoxycholate) is used as a detergent. Aggregation number 3–12; CMC 2–6 mm.

Deoxycorticosterone

Abbreviation used: DOC. DOC is Kendall's desoxy compound B, Reichstein's substance Q, 21-hydroxyprogesterone, 21-hydroxypregn-4-ene-3,20-dione. It is a mineralocorticoid hormone having a little glucocorticoiddactivity and is synthesized by the adrenal gland. It can be considered as its ester, deoxycorticosterone acetate (Abbreviation used: doca or DOCA) for clinical purposes.

Deoxyribonuclease

It is pyrimidine dimer. This has another name: endodeoxyribonuclease (pyrimidine dimer). Deoxyribonuclease is an enzyme that is involved in DNA repair. Furthermore, it can induce single-strand breaks in DNA on the 5′- side of pyrimidine dimers in the strand that contains the lesion, after the formation of the dimersfbecause of ultraviolet radiation.

Deoxyribonucleoside Monophosphate Kinase

These are any of several enzymes within sub - subclass EC 2.7.4 that catalyze the transfer of a phosphoric residue from a nucleoside triphosphate to a deoxyribonucleoside monophosphate, for example, from ATP to a deoxyribonucleoside phosphate in order to form ADP and a deoxyribonucleoside diphosphate.

Dermaseptin

Dermaseptin is referred to as any of a series of antimicrobial peptides which are derived from amphibian skin. Dermaseptins S1–5 constitute a family of cationic lysine - rich amphipathic antifungal peptides of 28 – 34 residues. It is pre - assumed to protect the naked frog skin from infections. Along with this, they have a lot of applications, and are used as the first vertebrate peptides in order to show lethal effects against the filamentous fungi that can cause infections in immunodeficiency syndrome, or during the use of immuno-suppressive agents. Dermaseptin is further classified in the TC system under number 1.C.52.

Dermatosparaxis

Dermatosparaxis is a disorder of cattle, sheep, cats, and dogs and is characterized by generalized skin fragility, early death from sepsis and joint laxity. In cattle, it is caused due to inactivating mutations in procollagen N-endopeptidase.

Descarboxy-Clotting Factor

This is referred to as any of the abnormal blood clotting factors that contain glutamic- rather than c-carboxyglutamic-acid residues. They are formed in animals that have vitamin K deficiency, or during the administration of vitamin K antagonists.

Desferrioxamine

This is a siderophore of the hydroxamate type which is derived from *Streptomyces pilosus* that tends to form a chelate complex specifically with Fe (III) ions, yielding ferrioxamine. Mainly, it is used for treating acute iron poisoning and for reversing iron and aluminum overload. The natural function that it contains is it transports iron into the microbial cell and/or also, aids in making iron available for the synthesis of heme.

Desmosome Or MaculaAdherens

Desmosome, commonly referred to as macula adherens, is a patch - like inter - cellular junction which is found in vertebrate tissue. It contains parallel zones of two cell membranes, which are separated by an inter - space of approximately 25–35 nmoand, have dense fibrillar plaques in the subjacent cytoplasm.

The inter - space is continuous with the inter - cellular space and moreover, contains a dense, central plaque of material that is rich in protein. Such types of proteins are known as linker proteins and further, consist of cell-cell adhesion molecules of the cadherin type.

Desmosomes are said to be important in cell - to - cell adhesion and are specifically numerous in stratified squamous epithelium which is subject to mechanical stress. The zona adherens is kind of similar to the inter - cellularhjunction, which is sometimes also referred to as a desmosome, which holdss the cell like a belt.

Diafiltration

Diafiltration is said to be the process of separating micro - solutes, for example, salts, which are from a solution of larger molecules (or of exchanging them for different micro - solutes) and separated by ultrafiltration after continuous addition of solvent (or of a solution of the new micro - solutes). This in turn removes the original micro - solutes rapidly from the solution, the volume of which remains constant.

Diagonal Chromatography

Diagonal chromatography is a two - dimensional chromatographic technique which is used in order to determine the sensitivity of a constituent present in a mixture to some (photo)chemical process, such as oxidation. The chromatography of the sample is done in one direction, and this process is carried out in situ. Further, the specimen is re - chromatographed at right angles. The compounds which are left unmodified by the treatment, tend to all lie on a diagonal line across the chromatogram.

Dialysable

Dialyzable means the capability to get purified or separated by dialysis. The term has also been used in order to describe material that either can or cannot traverse a dialysis membrane. In certain cases, the preferree term is diffusible for materials that can traverse a dialysis membrane.

DiaminobenzoicAcid

Abbreviation: DABA, is also referred to as 3,5 - diaminobenzoic acid. It is a compound that provides fluorescent products when it is heated in mineral acid solution along with aldehydes. Further, it is used for the microfluorimetric determination of DNA (the reason being that it does not react with RNA) and in the analysis of sialic acid.

Diauxy

Diauxy can be defined as the adaptation of microorganisms to culture media that contains two different carbohydrates. The growth occurs in two phases which are separated by a period of less rapid or even zero growth. Considering the first phase, the organism tends to utilize the carbohydrate for which it possesses constitutive enzymes. During the interlude, the organism tends to synthesize the required inducedsenzymes, and this is done for the metabolism of the other carbohydrate.sThese continue to develop in the second growth phase.

Dicer

Dicer is referred to as a multidomain endonuclease of ribonuclease III-type which is involved in the generation of siRNA that is small interfering RNAs and other forms of miRNA i.e., microRNA species. In *Caenorhabditis elegans*, it is important for normal development. The human and *Drosophila* enzymes are homologous and moreover, contain a DEXH - box ATP - dependent RNA helicase domain, tandem RNase III motifs, and a C - terminal dsRNA - binding domain. Considering animals, Dicer and its cofactors are not present in differentiated cells.

Dideoxynucleoside Triphosphate

Abbreviation: ddNTP, is referred to any artificial nucleoside triphosphate in which hydrogen atoms replace both of the hydroxyl groups on C2′ and C-3′ of the pentose moiety. They are used to sequence the DNA molecules by the chain - termination method.

Difference Spectrophotometry

It is a spectrophotometric method in order to investigate the effects of potential perturbants on a chemical substance or system in the solution. There are two samples that are further prepared and that contain identical solutions except that one which contains a potential perturbant, for example, a protein denaturant. At each frequency, the absorbances of the two samples are subtracted one from another which in turn generates a different spectrum and further, it highlights any small differences between the spectra of the normal and perturbed systems.

Differential Thermal Analysis

DTA is referred as differential thermal analysis which is a method in which a temperature of a sample is analyzeed in transitions. In this, the sample and an inert reference material are either heated or cooled at the same ratio and the difference in temperature between them is constantly monitored. The difference is constant or zero until or unless a thermally induced transition happens in the sample, when there is a change in the difference in temperature. The direction of the change signifies whether the transition is exothermic or endothermic.

Diphenylamine Reaction

In this reaction when deoxypentoses react with the acidic solution of diphenylamine, it gives a blue color. This reaction is used in the quantitative determination of DNA, and in combination or alone with other reagents as a detection reagent.

Diplochromosome

Diplochromosome is referred as a "double chromosome", it is comprising the arms of two daughter chromosomes that are attached to a single centromere. Diplochromosomes arise because of the failure of centromerr to divide following the duplication of the chromosome.

Dismembrator

Dismembrator is used as an ultrasonic device that allows rapid disrupting cells in cytological and bacteriological studies. This device consists of a probe which is oscillating at ≈20 kHz and which has an acoustical power that reaches up to 150 W and is dipped into a cell suspension, which is either contained in either a glass beaker or in a stainless-steel beaker.

Displacement Chromatography

This is defined as an analytical chromatographic technique. In this, a mixture is applied to a chromatographic column and then, the components of the mixture are successively displaced by elution with a solution. This solution contains another substance of higher affinity for the column material in comparison to that of the component which is the most firmly held component. The relative proportions of the components present in the mixture are determined by performing the frontal analysis of the eluate.

Distillation

Distillation can be defined as the act or process of evaporating a liquid by boiling. This in turn separates the components which are vaporized at different temperatures or at different rates. Further, the components condense back to a liquid (or solid). It is used for purifying one liquid from a mixture of liquids or from (dissolved) solids.

Figure 10. Chemistry Distillation Osmosis.

Source : Image By Pixabay

DNA Photolyase

EC 4.1.99.3; The recommended name for DNA photolyase is deoxyribodipyrimidine photolyase. Another name of the same is photoreactivating enzyme. This is generally an enzyme that splits cyclobutadipyrimidine (in DNA) into two pyrimidine residues (in DNA). Aand this helps in repairing the DNA after light - induced pyrimidine dimer formation. It is dependent ondlight and requires the co - factors to have reduced flavin and either (in some other different enzyme types) a folate coenzyme or even modified flavin.

DNA Sequencing

DNA sequencing determines the order in which deoxynucleotides occur in DNA. The fragments present in DNA, most of the time which are cloned in a plasmid vector, are subjected to either the chemical cleavage method or the chain - termination method, which is much more commonly used.

Dolichol Kinase

EC 2.7.1.108; It is generally an enzyme that tends to catalyze the formation of dolichyl phosphate from CTP and dolichol with the release of CDP. Dolichyl phosphate is referred to as a precursor of dolichylphosphoglucose, dolichyldiphosphooligosaccharides, and dolichylphosphomannose, which are the intermediates in N - linked glycoprotein oligosaccharide biosynthesis.

Dopamine

Dopamine is the alternative name for 3,4-dihydroxyphenylethylamine; 3-hydroxytyramine; 3,4-dihydroxyphenylethylamine; 4-(2-aminoethyl)-1,2-benzenediol. It is a catecholamine neurotransmitter which is formed by aromatic - L - amino - acid decarboxylase, EC 4.1.1.28, from 3,4-dihydroxy-L-phenylalanine. This is also referred to as a metabolic precursor of norepinephrine and epinephrine, which is found in dopaminergic nerves in the brain and in the adrenal medulla.

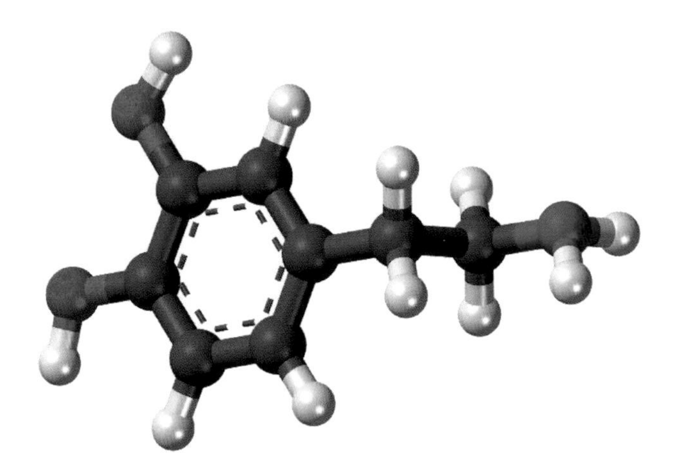

Figure 11. Ball-And-Stick Model Of The Dopamine Molecule.

Source : Image By Wikimedia Commons

Dosage Compensation

Dosage compensation is the epigenetic inactivation of one of the two X chromosomes of female cells, which leads to them having only one functional copy. This in turn makes them comparable with X - chromosome dosage in male cells, which have only one.

E

Eadie–Hofstee Plot

This plot refers to a graphical representation of enzyme kinetic data. In this plot the reaction velocity is divided by the substrate concentration, that is v/[S]. Eadie-Hofstee plot is plotted as ordinate against the reaction velocity, v, as abscissa. If in one case, a straight line is obtained, then the intercept on the abscissa gives the maximum velocity, V, and that on the ordinate V/Km. Therefore, the slope becomes –1/Km.

EaeI

EaeI is referred to as a type 2 restriction endonuclease; the recognition sequence is: [TC]↑GGCC[AG]. CfrI is an isoschizomer.

Eagle's Medium

Any of the different growth or maintenance media that are used in tissue culture and necessarily containing a balanced salt solution is referred to as Eagle's medium. Usually, the balanced salt solution is Earle's balanced salt solution or Hanks's balanced salt solution, which is supplemented with amino acids, serums, vitamins, and antibiotics.

Eam1105I

This is a type 2 restriction endonuclease; the recognition sequence is: GACNNN↑NNGTC.

Earle's Balanced Salt Solution

Earle's Balanced Salt Solution can be defined as a solution that is used in tissue culture in order to provide a normal ionic, pH, and osmotic environment for cell growth. Moreover, it contains, per 100 ml distilled water, 0.04 g KCl, 0.68 g NaCl, 0.02 g $CaCl_2$, 0.02 g $MgSO_4.7H_2O$, 0.0125 g NaH_2PO_4 $.H_2O$, 0.22 g $NaHCO_3$, 0.1 g glucose, and 0.002 g phenol red. Also, it contains a pH of 7.6 – 7.8.

Early Enzyme

It refers to an enzyme that is transcribed from an early gene of a virus. Any of the viral genes that is transcribed much early after the virus tends to infect a host cell. Such kinds of genes probably code for proteins that are required for viral nucleic acid replication.

Early Gene or Immediate Gene

Immediate gene is defi number of genes that are involved in the earliest responses of cells to factors and this further initiates the transition from quiescence to proliferation. The early genes include those which tend to encode transcription factors, including the c-jun, c-fos, and c-e, and alsoand, such coding for structural proteins like actin. Further, these early genes are sub - divided into immediate - early and delayed - early genes.

Early Growth Response Protein

It is any of a family of proteins that are expressed early in response to growth factors and moreover, function as transcriptional regulators, which bind to the DNA sequence 5′-CGCCCCCGC-3′.

Early Protein

Early protein is a protein that is transcribed from an early gene of a virus. The early quitter is any incomplete polypeptide that is formed in an *in vitro* translation system.

Easson–Stedman Model

Easson-Stedman model can be referred to as a model which explains the differences in biological activity considering two enantiomeric molecules. Furthermore, it assumes that three of the groups which are attached to a carbon atom which is asymmetric in an agonist (or substrate) are involved in its attachment to a particular receptor (or enzyme), and this is the reason that one enantiomer will be a better agonist (or substrate) than the other.

4E-BP Or Eif4e-Binding Protein

It is s any of a group of small proteins which bind eIF4E (the mRNA 5′ cap-binding protein) and further, in turn, controls its availability for the purpose of initiating translation. The state of phosphorylation of 4E-BP tends to control its binding to eIF4E. Under particular conditions which induce apoptosis, the interaction of 4E-BP with eIF4E further increases, and then there is a decrease in the initiation of translation.

EC50

EC50 refers to half effective concentration. It can further be explained as the molarity of an agonist which tends to produce 50% of the maximal possible effect of that agonist. Other percentage values including EC20, EC40, etc. can thus be specified. The action of the agonist can be either stimulatory or inhibitory.

Eccrine

Eccrine describes a secretory cell that discharges its product without losing the cytoplasm. Moreover, it describes a gland that is made up of such cells.

Ecdysis

Ecdysis is the periodic shedding of the exoskeleton as well as the construction of a new cuticle (as in insects and crustaceans). It can also be referred to as the shedding of the outer layer of the skin, similar to that which happens in reptiles

Ecdysone

Alphaa-ecdysone (22R)-2b,3b,14,22,25-pentahydroxycholest7-en-6-one; It is the ecdysteroid that is synthesized in and further secreted by the prothoracic glands of immature insects and the ovaries of the adult females. Firstly, it is isolated from the silkworm moth (named as Bombyx mori), it is the inactive pro - hormone of the moulting hormone ecdysterone. In addition to this, it can have intrinsic hormonal activity at some other stages of insect development.

Ecdysone

Ecdysone can be defined as inducible mammalian expression system. This is a system for the expression of exogenous transgenes in mammalian cells. In such cells, the timing of expression tends to depend on the administration of the insect hormone ecdysone.

Ecdysone Receptor

It is a receptor for ecdysone Ecdysone receptor is located in the cell nucleus and is related to the steroid/thyroid/retinoic acid family of nuclear hormone receptors. Further, it contains a heterodimer of a sub - unit named EcR i.e., ecdysone receptor and another named USP (that is ultraspiracle), the insect homolog of vertebrate retinoid X receptor. The heterodimer further binds ecdysone with high affinity, and the ternary complex binds to DNA.

E-Cell

E-cell refers to a computer model of a cell that is generated using proteomic, genomic, transcriptomic, and metabolomic data.

Echinocandin

It is any of a group of cyclic hexapeptides that consists of an N-acyl aliphatic or aryl side chain. They inhibit the synthesis of 1,3bglucans of fungal cell walls and are said to be effective against *Candida* spp., *Aspergillus* spp., and *Pneumocystis carinii*. Some of the examples include micafungin, caspofungin, and anidulafungin.

Echinochrome A

It can be elaborated as 2-ethyl-3,5,6,7,8-pentahydroxy-1,4-naphthalenedione; Echinochrome A is a red pigment of sea urchin eggs.

Echinoderm

This refers to any of the phylum Echinodermata of marine invertebrates which consists of sea urchins, starfish, sea cucumers, odiesy arssradially calsymmetrical radially, having calcareous skeletal plates in their skin and also, a well - developed coelom. By means of retractable tube feet, locomotion and gaseous ecarried outaccomplished. They are further related to the ancestral speciesstock which gave rise to the vertebrates.

Eclosion Hormone

It refers to any peptide hormone which tends to program the death of certain muscles and neurons during the process of metamorphosis of insects.

Eco52I

It refers to a type 2 restriction endonuclease. Its recognition sequence is: C↑GGCCG. XmaIII is an isoschizomer.

Eco81I

Eco81I is a type 2 restriction endonuclease. The recognition sequence is: CC↑TNAGG. SauI is an isoschizomer.

Ecology

Ecology can be defined as the branch of biology that deals with the relation between living organisms and their surroundings.

Figure 12. Ecology World Summer.

Source : Image By Pixabay

Ecomone

Ecomone is any nontrophic molecule that makes sure of a flux of information between organisms in an ecosystem.

EcoO65I

EcoO65I is a type 2 restriction endonuclease. The recognition sequence is: G↑GTNACC. BstEII and BstPI are isoschizomers.

EcoO109I

It is a type 2 restriction endonuclease. The recognition sequence is: [AG] G↑GNCC[TC]. DraII is an isoschizomer.

EcoRI

EcoRI is a type 2 restriction endonuclease. The recognition sequence is: G↑AATTC. The M.EcoRI modification site is A3.

EcoRV

This is a type 2 restriction endonuclease. The recognition sequence is: GAT↑ATC.

Ecosystem

An ecosystem can be defined as a unit of the environment together having the organisms that it has. There refers to a constant interchange between living organisms and their chemical as well as the physical environment.

EcoT14I

It is a type 2 restriction endonuclease. The recognition sequence is: C↑C[AT] [AT]GG. StyI is an isoschizomer.

EcoT22I

This is a type 2 restriction endonuclease. The recognition sequence is: ATGCA↑T. AvaIII is an isoschizomer.

Ecotin

Ecotin refers to a monomeric periplasmic protein of *Escherichia coli* that tends to inhibit the pancreatic serine proteases chymotrypsin, trypsin, and elastase. Further, it allows the organism to survive being in the presence of these enzymes. In addition to this, it inhibits blood coagulation factors Xa , and kallikrein, and alsoand, has anticoagulant effects.

ECTEOLA Cellulose

ECTEOLA cellulose can be defined as a weakly basic (pK ≈7.5) anion - exchange material of uncertain structure, which is prepared by the condensation of epichlorhydrin, triethanolamine (hence e,c + t,e,ola), and cellulose. It is said to be useful in separations of proteins, nucleoproteins, and nucleic acids.

Ectodysplasin A or EDA-A

EDA-A is a type II transmembrane protein of the TNF that is tumor necrosis factor family which includes a furin recognition sequence. It is present in hair follicles, keratinocytes, and sweat glands. Furthermore, it regulates the development and differentiation of HO HO OH O OH O OH CH3 epidermal structures. There are two iso-forms: EDA-A1 (391 amino acids) and EDA-A2 (389 amino acids). Their receptors tend to activate NF-jB. Mutations in the furin recognition sequence further cause X-linked anhidrotic ectodermal dysplasia.

Ectoenzyme

Ectoenzyme is an enzyme that is attached to the external surface of the plasma membrane of a cell.

Ectohormone

Ectohormone is an ectocrine substance that is when produced or released, benefits either the organism which is producingtit, or other members contained in the same species.

Ectoparasite

Ectoparasite refers to any parasite that lives on the exterior of its host organism.

Figure 13. Norwegian Soft Tick.

Source : Image By Flickr

Ectopic

This means the phenomena of occurring in an unusual place or in an unusual manner or form. Let's take an example. An ectopic protein is referred to as a protein that is produced by a neoplasm derived from tissue. This tissue does not normally produce that specific protein. Further, an example is the production of vasopressin and corticotropin by small cell lung cancer. Another example includes ectopic pregnancy which tends to occur when the fertilized ovum implants outside the uterus in a fallopian tube.

Ectoplasm

Ectoplasm refers to the outer part of the cytoplasm which is relatively rigid.

Ectoplast

Ectoplast can be defined as the part of the plasma membrane of a plant cell that is in contact with the cell wall. The ectoprotein refers to any individual protein which is found on the exterior of cells. Such kinds of proteins can function as mediators of cell – cell or cell – surface interactions (for example), or as receptors for substances performing regulatory actions on cells.

Ectosymbiont

Ectosymbiont is a partner present in a symbiotic relationship which remains outside the tissues and cells of the other partner. Most of the time, it occupies a body cavity; some examples include one of the cellulose - metabolizing microorganisms which occur in the digestive tract of ruminants.

ED50

ED50 refers to the median effective dose. This means the dose of a drug or other agent which produces, on average, a particular all - or - none response in 50% of a test population. If in case, the response is graded, the dose which produces 50% of the maximal response to that drug or agent.

Edeine

Edeine refers to the pentapeptide-amide antibiotics which are produced by a strain of *Bacillus brevis* and are said to be effective against Gram - positive and Gram - negative bacteria, some fungi and some other eukaryotic cells, and some neoplastic cells.

EDEM

EDEM refers to a stress - inducible ER that is an endoplasmic reticulum membrane protein that is homologous with a - mannosidase but is inactive catalytically. It further interacts with calnexin and tends to recognize the folding status of glycoproteins in the ER, through its trans - membrane segment.

Edema

Edema refers to the swelling of an organ or tissue because of the accumulation of fluid.

Figure 14. Catastrophic Edema Of The Legs.

Source : Image By Wikimedia Commons

Edema Factor

Edema factor can be defined as a protein component of the *Bacillus anthracis* toxin (which is responsible for anthrax) and tends to cause tissue swelling.

Edestin

Edestin refers to a 300 kDa globulin that is obtained from hemp seed, castor-oil beans, and some other seeds. Moreover, it readily forms polymorphic crystals and supports the growth ofnanimals when other dietary proteins are not present.

EDG Or Endothelial Differentiation Gene

EDG is Endothelial Differentiation Gene refers to a gene that encodes a G - protein - coupled receptor which is present on endothelial cells. For the receptor, a major ligand is sphingosine 1 - phosphate, which when bounded tends to activate nitric oxide synthase.

Editosome

Editosome refers to a macromolecular complex which is involved in the editing of RNA transcripts. Particular ribonucleotide complexes include guide RNA, which specifies the edited sequence.

Edman Degradation

Edman degradation can be defined as a procedure that is used in sequencing (poly)peptides. In this process, amino - acid residues are removed step - by - step, from the N terminus by reacting it with phenylisothiocyanate.

EDTA

EDTA refers to ethylenediaminetetraacetate; edetate. EDTA is any anion or salt, especially the (hydrated) disodium salt, of edetic acid, ethylenediamine N, N, N′, N′-tetraacetic acid, (ethylenedinitrilo) tetraacetic acid. It is commonly used as disodium salt. It is considered to be a powerful chelating agent for divalent metal ions.

Figure 15. Sample Of Ethylenediaminetetraacetic Acid Disodium Salt.

Source : Image By Wikimedia Commons

EEDQ

EEDQ refers to N-ethoxycarbonyl-2-ethoxy-1,2-dihydroquinoline. It is an agent for peptide condensation which causes little or no racemization. Moreover, it is useful in coupling ligands to insoluble polymers. Also, it causes progressive as well as irreversible inhibition of mitochondrial F1 - ATPase.

EFEMP

EFEMP refers to EGF - containing fibrillin - like extracellular matrix protein. Consider the two glycoproteins, either of the two contains multiple EGF further repeats and belong to the fibrillin family of extracellular matrix proteins. A missense mutation in the EFEMP1 gene is considered common in an autosomal dominant form of retinal dystrophy. EFEMP2 is further encoded at a different locus.

Effective Theoretical Plate Number Symbol: N

It refers to a number indicative of chromatographic column performance when resolution is considered.

$N = 16Rs\ 2 /(1-a)\ 2$,

where Rs is the peak resolution and a is the separation factor

The separation factor is defined as the ratio of the distribution ratios or coefficients for the two substances which are being resolved when measured under identical conditions. Considering the definition, a is greater than unity. The comparison must be done between height equivalent to an effective theoretical plate.

Effectomer

Effectomer can be defined as the part of a two - component agonist which in turn brings about the biological effect. This can only be done when it is bound to a haptomer. Haptomer enables it for interacting with the cell membrane and further, exert its toxic action on intact cells. An example can be taken of the A fragment of diphtheria toxin that can only exert its toxic effects on a target cell when it is bound to the B fragment.

- 2766633126 need to check smilekids dentistry
- 7034351500 no dmo (pediatric dentistry)
- 5716858039 go smiles no dmo new patient 249 dollars
- 5714975812 closed hero pediatric dentistry
- 7034294619 closed dentistry of nothen va

Effector

Effector can also be referred to as modifier or modulator. Effector is any small molecule or ligand which interacts with an enzyme further changing its catalytic behavior. During the enzyme action, it does not get changed itself. A positive effector tends to enhance catalytic activity whereas on the other hand there is a reduction due to a negative effector. In addition to this, the effector is a cell or organ which produces a physiological response when stimulated by the nervous system.

Effector Site

The effector site is a regulatory site. It can be defined as any site on an enzyme molecule that binds an effector.

Efficacy Symbol

Efficacy symbol, represented by e, can be defined as a quantitative index of drug action related to the magnitude of tissue response. This tissue response is generated at a given level of receptor occupancy with (full) agonists (having high efficacy) which in turn generates the maximum possible response.

Drugs, having lower efficacy, which cannot generate a maximal response even in the situation when all receptors are occupied are referred to as partial agonists. The antagonists have an efficacy of zero.

Efficiency

Efficiency can be defined as the ratio of the useful output of a machine, system, device, etc. to the input. It can be in terms of mechanical, thermal, or radiation energy, or of biological or chemical conversions.

EOP

EOP is efficiency of plating, can be defined as a quantification of the relative efficiencies due to which different cells can be infected by viruses, and moreover, support viral replication. It refers to the ratio of the plaque count to the number of virions in the inoculum.

EF-G Or Translocase

EF-G is an elongation factor of *Escherichia coli* that promotes the translocation, which is GTP - dependent, of the nascent protein chain from the A - site to the P - site of the ribosome. In eukaryotes, an equivalent factor is named eEF-2.

EF-Hand

EF-hand refers to a helix - turn - helix motif which binds Ca^{2+}. It is found in a number of Ca^{2+}-binding membrane proteins. For example, in some others, the myosin light chain involves helices E and F of the Ca^{2+}-binding protein.

Efrapeptin Or Efrastatin

Efrapeptin or efrastatin refers to a hydrophobic peptide fungal antibiotic that inhibits oxidative phosphorylation. This is done by binding to the soluble component, F1, of mitochondrial H+ - transporting ATP synthase.

Egasyn

Egasyn is an amphipathic glycoprotein that is found in microsomal (that is endoplasmic reticulum) membranes of mouse liver. This tends to form a complex with microsomal (but not lysosomal) b-D-glucuronidase. Further, anchors it to the microsomal membranes. In a similar manner, it may anchor other polar proteins. It tends to hydrolyze carboxylic esters to an alcohol and carboxylic anion. It contains a C - terminal consensus sequence, HXEL, which further retains the proteins in the ER that is endoplasmic reticulum.

EG Cell

EG cell is also referred to as L cell. It is an enteroglucagon - producing endocrine cell. Such types of cells are found in the basal part of the mucosal glands present in the lower intestine. The highest concentrations tend to occur in the terminal ileum and colon. They consist of dense, large, secretory granules and are easily differentiable from other endocrine cells in the same area.

Egg

An egg is defined as the reproductive structure of certain animals including reptiles, birds, and insects. They consist of an ovum along with nutritive and protective tissues. From this, after the fertilization, a young offspring emerges.

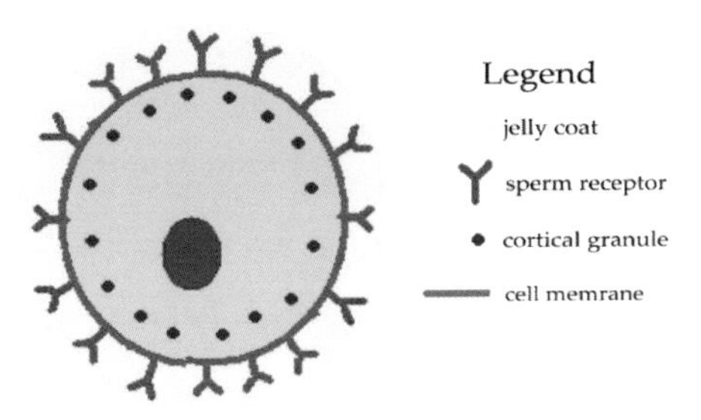

Figure 16. Components of A Human Egg Cell Prior To The Process Of Fertilization.

Source : Image By Wikimedia Commons

EGR2

EGR2 refers to early growth response 2. It is a C2H2 - type of zinc - finger protein that is a transcription factor that is involved in myelination in the peripheral nervous system.

F

F +

It is a symbol that refers to a donor bacterial cell that contains an F plasmid.

F–

It is a symbol that denotes a recipient bacterial cell that lacks an F plasmid.

F9

This refers to a continuous line of mouse teratocarcinoma cells that retain the capacity for differentiation. Moreover, they can give rise to chimeric mice when they are injected into the blastocoel of mouse blastocysts. They are then allowed to complete the developmental process in utero.

Fab Fragment

Fab fragment is a ≈45 kDa protein fragment which is obtained (together with Fc fragment and Fc ′ fragment) by the process of papain hydrolysis of an immunoglobulin molecule. It contains one intact light chain which is linked by a disulfide bond to the N - terminal part of the contiguous heavy chain (the Fd fragment). From each IgG antibody molecule, two Fab fragments are obtained; each and every fragment contains one antigen - binding site.

F(Ab')2 Fragment

This is a ≈90 kDa protein fragment which is obtained (together with pFc ′ fragment) by performing the process of pepsin hydrolysis of an immunoglobulin molecule. It contains that part of the immunoglobulin molecule N terminal to the site of pepsin attack and further, contains both

Fab fragments that are held together by disulfide bonds in a short section of the Fc fragment (the hinge region). From each IgG antibody molecule, one F(ab')2 fragment is obtained; it contains two antigen - binding sites but not the site for complement fixation.

FABP

FABP refers to Fatty Acid Binding Protein. In plasma, albumin is the major FABP, whereas a 40 kDa plasma membrane FABP of the liver is considered to be involved in the uptake of oleate in an ATP- and Na+-dependent manner. The small cytosolic FABPs (≈130 amino acids) have been characterized from liver, heartalintestine, skeletal muscle, and some other tissues. The crystal structure of FABP from rat intestine further shows 10 antiparallel beta-strands in two sheets that form a hydrophobic binding pocket; four motifs are characteristic of this function.

Fabry's Disease

Fabry's disease is a disease which is an X - linked inborn error of human metabolism because of defective lysosomal a - D - galactosidase A activity. Moreover, it is a sphingolipidosis or sphingolipid lysosomal storage disease. The enzyme defect results in the progressive deposition in most visceral tissues of neutral glycosphingolipids along with terminal alpha- D - galactosyl residues, principally globotriaosylceramide.

Facilitated Diffusion

It can be defined as a diffusion that occurs across a biological membrane. This diffusion takes place through the participation of particular transporting agents (transporters) or carriers. The equilibrium distribution achieved is similar to that which is achieved by simple diffusion. But at any one site, the facilitated diffusion is mediated by a transport protein which in turn exhibits specificity for the transported species. The existence of the transporter tends to mediate passage across the membrane of molecules to which it would otherwise be impermeable.

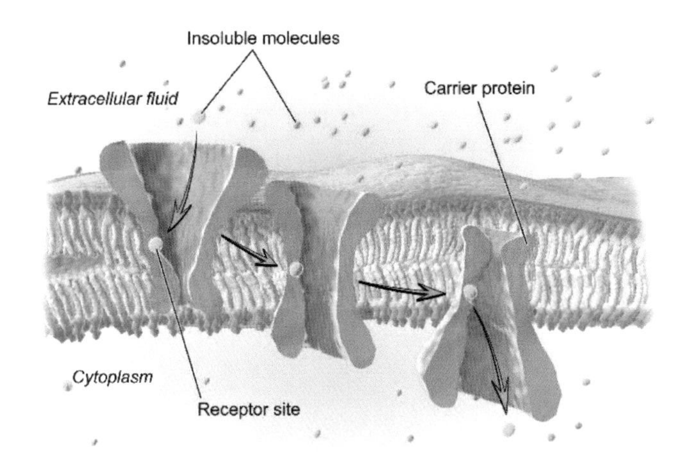

Figure 17. Facilitated Diffusion.

Source : Image By Wikimedia Commons

FACIT

FACIT refers to Fibril - Associated Collagens with Interrupted Triplex helice. FACIT is a group of collagens (types IX, XII and XIV) composed of triple - helical regions that are interrupted by non - triple - helical regions of 8 – 42 amino acids.

Factor

A factor can be defined as any componese which contributes to an effect or result. Most of the time, this term is used to denote an uncharacterized or incompletely characterized component of a biological system.

Factor F430

It is a tetrapyrrole compound that contains nickel. It is the prosthetic group of EC 2.8.4.1 coenzyme B sulfoethylthiotransferase (methyl coenzyme M reductase). This in turn catalyzes the terminal reductive step of methanogenesis in methanogenic bacteria, in which methane is formed from methyl - coenzyme M.

Factor S

Factor S is a tetrapeptide (of unknown structure) that is isolated from the cerebrospinal fluid of goats. This slowly induces delta (that is slow wave) sleep in rats on infusion into the cerebral ventricles.

Facultative

Facultative means one's ability to live under more than one set of environmental conditions.

FADD

FADD is the abbreviation for Fas - Associated protein with Death Domain. Another name for FADD is MORT1. This is a cytosolic adaptor protein that binds Fas (via the death domain on each) and procaspase 8 (via the death effector domain on each). The reshus formed is a complex form that s part of the DISC that is Death - Inducing Signalling Complex.

Fahrenheit Temperature Scale

Fahrenheit Temperature Scale can be defined as a temperature scale in which 32°F is set equal to 273.15 K. And 212 °F is set equal to 373.15 K. Therefore, 1 Fahrenheit degree becomes equal to 5/9 × K, and Fahrenheit temperature (°F) = 9/5 × (Celsius temperature) + 32.

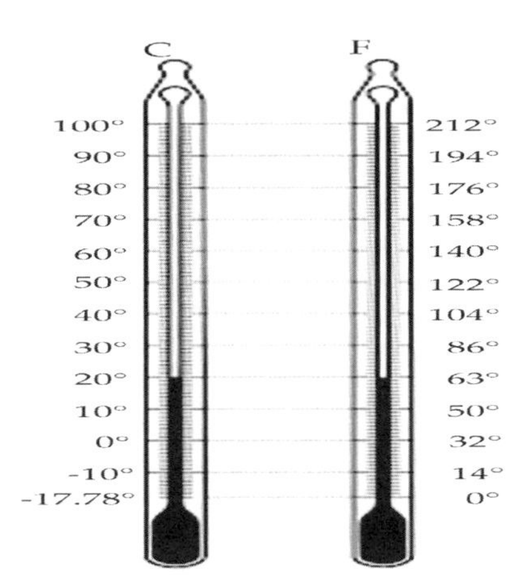

Figure 18. Comparison Of Fahrenheit And Celsius Temperature Scales.

Source : Image By Wikimedia Commons

Falcipain

Falcipain refers to a cysteine protease of *Plasmodium falciparum*. Further, it may be involved in host cell invasion by the parasite.

Fall Curve

Fall curve can be defined as the curve which describes the reduction in the color intensity of a sample with time which is obtained in a colorimetric analysis by using an autoanalyzer.

Falling – Drop Method

Falling - drop method is a method rder to determine relative densities of liquids accurately. In this method, a drop of the liquid is made to fall to its equilibrium position in a column which contains a defined density gradient of an immiscible liquid. Formerly, it was used in order to determine the deuterium oxide content of samples of water.

Falling – Sphere Viscometer

Falling - Sphere Viscometer can be defined as an insorder toto measure the viscosity of a liquid from the time taken for a solid sphere to fall through a given distance in a column of the liquid in comparison to the time taken for it to fall through the same distance in a liquid of given viscosity.

Familial Hypercholesterolemia

This is a common autosomal dominant genetic condition that results in elevated levels of low - density lipoprotein in blood. The fatty deposits further accumulate in the arteries and skin, and the condition predisposes to coronary heart disease. This is a result of mutation of the low - density lipoprotein receptor gene leading to haploinsufficiency in the heterozygous state.

Familial Hyperlysinemia

This is an apparently benign genetic human disorder in which decreased activity of the bifunctional enzyme alpha-aminoadipic semialdehyde synthase results in hyperlysinemia and hyperlysinuria.

Familial Mediterranean Fever

Familial Mediterranean Fever refers to an autosomal recessive disorder characterized by periodic fever as well as systemic amyloidosis with renal involvement. It is prominent in Sephardic Jews and Armenians, and further, it is caused by a lot of mutations in the gene (locus at 16p13) for a protein (781 amino acids) of unknown function known as pyrin or marenostrin.

FAN

FAN refers to Factor Associated with Neutral sphingomyelinase activation. This is a protein of the WD40 family that couples the p55 TNF receptor and the cannabinoid receptor to neutral sphin in order toto release ceramide as a signaling molecule from sphingomyelin on the inner leaflet of plasma membranes.

Fanconi Anemia

Fanconi anemia is a rare autosomal recessive and highly variable disorder. This is characterized by bone marrow failure, increased susceptibility to cancer and developmental abnormalities. Usually, it proves fatal by the second decade of life. The cells from patients further show increased susceptibility to cross - linking (alkylating) agents. The eight complementation groups are recognized: A, B, C, D1, D2, E, F, and G. Mutations in any of these gene products results in the disease.

Fanconi Syndrome

Fanconi syndrome can be defined as a human disorder that is characterized by generalized dysfunction of proximal renal tubules. It is generally manifested as glycosuria, amino acidurihaturia, andphosphaturia, and vitamin D - resistant bone disease. Moreover, in adults, it is manifested as rickets in children or osteomalacia. It can be acquired as result of different toxic agents (for example, heavy metals or some specific drugs) or be related to a variety of inherited metabolic disorders.

Faraday Constant Symbol: F

This refers to a fundamental constant. Faraday constant symbol can be defined as the quantity of electricity that is required to deposit one mole of a univalent ion from a solution of an electrolyte.

Farnesoid X Receptor

FXR is Farnesoid X Receptor or bile acid receptor. FXR is an orphan receptor that is restricted to the enterohepatic system, kidney, and adrenal cortex. Further, it forms an obligate heterodimer with retinoid X receptor (which in turn binds 9-cis-retinoic acid), and the dimer furtcular hormonehormone response elements on DNA. Bile acids are generally the endogenous ligands for FXR.

Farnesol

Farnesol is the trivial name for any of four possible stereoisomers of 3, 7, 11 - trimethyl - 2, 6, 10 dodecatrien - 1 - ol; it is a sesquiterpene alcohol. The 2E, 6E - isomer is found in a number of essential oils. The 2Z, 6E - isomer is considered to be a minor constituent of some of the essential oils.

Farnesyl transferase

Farnesyl transferase refers to an enzyme that in turn catalyzes the formation of presqualene diphosphate from two molecules of farnesyl diphosphate along with release of pyrophosphate. Furthermore, it occurs in the facultative anaerobe farnesyltransferase pathway in order to synthesize the squalene and derivatives (cholesterol, sesquiterpenes). Considering eukaryotes, it can be said that it is a monomeric microsomal enzyme.

Farnesyl transtransferase

This is an enzyme that tends to catalyze the formation of geranylgeranyl diphosphate from trans, trans-farnesyl diphosphate and isopentenyl diphosphate along with the release of pyrophosphate. Moreover, it occurs in the pathway for sesquiterpene and cholesterol synthesis and tends to form the prenyl derivative geranylgeranyl diphosphate. Considering some of the fungi and plants, this kind of activity is part of a multifunctional protein that also has the activities of geranylgeranyl-diphosphate synthase.

Farr Test

Farr test is a radio - immunoassay teused in orderto measure the absolute antigen - binding capacity of an antiserum. The antibody is further allowed to react with radio - labeled antigen and the antigen – antibody complex is further precipitated. This precipitation is done by the addition of saturated ammonium sulfate solution to a final concentration of 40 w/v.

Far Western Analysis

Far Western analysis can be defined as a modification of Western blotting. In this process, one of a complex mixture of proteins, which is separated by either denaturing or non - denaturing gel electrophoresis, is further detected by interacting with a bait protein. The bait protein can provide an epihe fact that it can be readily detected with anti - tag antibody labeled with an enzyme.

Fascicle

This refers to a fiber tract. It is a tight parallel bundle of nerve fibers which are either axons or dendrites. Further, it is formed during the process of growing of the nerves. This is mediated in part by cell adhesion molecules.

Fasciclin

Fasciclin can be defined as any of the proteins of a related group. Such proteins are involved in nerve fascicles (bundles). They include fasciclin I and fasciclin II. Fasciclin I is neuronal cell adhesion molecule. Fasciclin II is a neuronal recognition molecule that is related to NCAM.

Fascin

Fascin can be defined as a protein that is involved in the formation of actin bundles and actin polymerization.

Fas Ligand

This refers to a homotrimeric protein on the cytotoxic T cell surface. This in turn binds with the trans -membrane receptor proteins on the target cell that is called as Fas. The binding alters the theFas proteins. The reason is that their clustered cytosolic tails recruit procaspase - 8, which results in apoptosis.

Fast Component

Fast component is an unusual component of a mixture, for example, a hemoglobin variant, which moves in a particular buffer system in a much more rapid way in chromatography or electrophoresis in comparison to the one which does the normal component.

Fast Reaction

Fast reaction refers to any reaction or any step in a reaction sequence. Such kind of reaction has a large rate constant and furthermore, proceeds in a rapid manner. Considering a reaction sequence, it can be said that a fast reaction is not the rate - limiting step.

Fat

Fat refers to any triacylglycerol or mixture of triacylglycerols. Fats are solid below 20°C, and the f . Those are liquid at such temperatures are usually considered as oils. Additionally, it is an alternative name for adipose tissue and an alternate name for lipids.

Fatal Familial Insomnia

Fatal familial insomnia is referred to as a rare inherited neurological disease. It was first described in the year 1986 among inhabitants of mountainous regions of Italy. Further, it is characterized by intractable insomnia and some other neurological abnormalities. It is considered to be one of three familial forms of human prion disease.

Fat Body

A fat body is any of the cellular structure which contains fat. This in turn serves as energy reserves in amphibians, insects, and in reptiles.

Fat Cell

Fat - cell or lipocyte or adipocyte (in animals) is considered as any living cell which contains noticeable amounts of lipid, primarily referred to as fat or oil.

Fat Index

The fat index can be defined as the mass of diethylether - extractable fat per unit mass of non-fat material in a foodstuff or tissue (for example).

F1-ATPase

F1-ATPase refers to the globular catalytic domain of the H+ - transporting ATP synthase.

Fo ATPase

Fo ATPase refers to the intrinsic domain of the H+ - transporting ATP synthase. The 'o' in the term generally refers to 'oligomycin - sensitive'.

Fat- Soluble Vitamin

Fat-soluble vitamin can be defined as any of a diverse group of vitamins that are soluble in organic solvents, but it is relatively insoluble in water. Such a group of vitamins includes vitamins A, D, E, and K.

Fatty Acid

Fatty acid is any of the aliphatic monocarboxylic acids which can be liberated by hydrolysis done from naturally occurring fats as well as oils. The fatty acids are predominantly straight - chain acids of 4 to 24 carbon atoms. These carbon atoms can be either saturated or unsaturated. Even the branched fatty acids and hydroxy fatty acids may also occur. Also, very long-chain acids of over 30 carbons are found in waxes.

Fatty-Acid Activation

Fatty - acid activation refers to the conversion of a fatty acid molecule to its fatty acyl - coenzyme A thioester is the first step considering the reactions of beta-oxidation.

Fatty-Acid Nomenclature

This refers to a system of symbols to describe the fatty acids. The basic symbolism contains the number of carbon atoms in the molecule. This is followed by the number of double bonds and these two numbers are further separated by a colon.

Fatty-Acid Oxidation Complex

This can be referred to as a bacterial multi-function protein complex which further catalyzes the beta-oxidation of fatty acids.

Fatty Acid Synthase Complex

Fatty acid synthase complex refers to a large multi - enzyme complex that in turn catalyzes a spiral set of reactions where the synthesis of fatty acids is done from one molecule of acetyl coenzyme A and successive molecules of malonyl coenzyme A.

Fatty-Acyl-CoA Dehydrogenase

Fatty-acyl-CoA dehydrogenase is referred to as any of the acyl-CoA dehydrogenase components of the beta - oxidation system. Further, the mammalian liver contains a family of these enzymes which are differentiated by their chain - lengths. Specifically, into very - long - chain (C20–C14), long - chain (C18–C12), medium - chain (C12–C4), and short - chain (C6–C4) acyl-CoA dehydrogenases.

Fatty-Acyl-CoA Synthase

This is the multi - functional protein that is responsible for synthesizing the long - chain fatty acyl moieties in some of the eukaryotes, for example, yeast.

Fatty Aldehyde: NAD+ Oxidoreductase

An NAD+-dependent long - chain - aldehyde dehydrogenase, EC 1.2.1.48, which tends to catalyze the conversion of fatty aldehydes (6C to 24C long) to the corresponding fatty acids. Further, it is anchored to the membrane of the endoplasmic reticulum through its C - terminal region. Also, it accompanies fatty alcohol: NAD+ oxidoreductase.

Fatty Degeneration

Fatty degradation is also referred to as fatty infiltration. This is the deterioration of any tissue because of the deposition of abnormally large amounts of fat, usually in the form of globules, in its cells.

Fatty Liver

Fatty liver refers to a pathological condition of liver tissue that has undergone fatty degeneration. It leads to the administration of different poisons, specifically chlorinated hydrocarbons. Fatty liver is generally a result of a dietary deficiency of either choline or threonine.

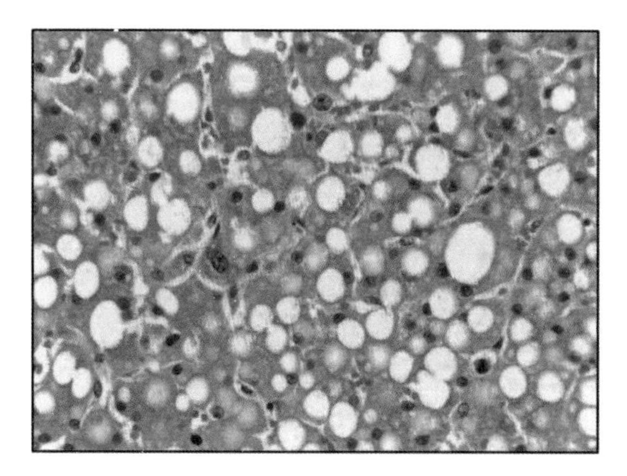

Figure 19. Fatty Change Liver: Hepatic Parenchymal Cell Cytoplasm Containing Clear Vacuoles Containing Fat Of Varying Sizes.

Source : Image By Wikimedia Commons

FATZ

FATZ refers to Filamin -, Actinin -, and Telethonin - binding protein of Z disc. It is any of three isoforms (such as FATZ1, 2, or 3) of a protein of the Z disc of skeletal muscle. It tends to bind c - filamin, a - actinin, telethonin, and calcineurin.

Favism

Favism is a disease of humans in which hemolysis is caused due to eating fava beans (*Vicia faba*) or by taking certain drugs. Some examples include sulfonamides, primaquine, and nitrofuran antibiotics. It is a basic defect which is an inherited deficiency of glucose - 6 - phosphate dehydrogenase in the red blood cells. Among them, about 130 variants are known.

F1 bacteriophage

F1 bacteriophage refers to a filamentous single-stranded DNA virus that tends to infect only male strains of *Escherichia cli.*

F1-CF0

F1 - CF0 is the e - sub-unit of the H+ - transporting ATP synthase complex in the chloroplasts of *Chlamydomonas reinhardtii.* In this, the subunitsidered to beis different from that in the higher plants.

G

GABAergic

A neuron or other cell that is affected by γ-aminobutyrate is said to be GABAergic. It also refers to nerves that function by releasing γ-aminobutyrate.

GABA-Modulin

GABA-modulin is a brain protein that inhibits the high-affinity GABA (c-amino-n-butyric acid) binding site on synaptic membranes.

GABA Receptor

GABA receptors are membrane proteins that bind the inhibitory neurotransmitter caminobutyrate (GABA) and influence its actions. GABAA receptors, also known as GABA-gated channels, are Cl– channels, whereas GABAB receptors are G-protein-coupled receptors.

GM2-Activator Protein

GM2-activator protein is a glycosylated protein (162 amino acids) that is required for lysosomal b-hexosaminidase A hydrolysis. It is encoded as a precursor by a gene on 5q32-q33. The AB type of Tay–Sachs disease is caused by a deficiency that can be caused by one of four mutations.

GADD

GADD is an acronym for growth arrest and DNA damage, and it refers to one of two proteins (GADD45 and GADD153) that are generated in response to oxidative stress, induce transient cell cycle arrest, and may play a role in DNA repair. p53 regulates the expression of the GADD45 gene.

Gaddum Equation

The Gaddum equation is a relationship that represents the scenario where two ligands, A and B, are in equilibrium with a shared binding site. It was proposed to replace the Hill–Langmuir equation. PAR = [A]/KA (1 + [B]/KB) + [A], where PAR is the fraction of ligand A's binding sites occupied; the square brackets denote concentrations.

GAGA Factor

GAGA factor is a *Drosophila* DNA-sequence-specific transcription factor that binds to numerous (GA) n repeats in the Hsp70 gene promoter. It works in tandem with NURF to cause nucleosome spacing in the promoter region to be disrupted.

GAL4

GAL4 is a gene involved in galactose utilization control in yeast. The GAL4 protein has a zinc-finger motif for binding to the UAS, which it subsequently activates, and a site for binding the GAL80 protein and galactose complex, which is encoded by the GAL80 gene. When this is bound, GAL4 protein activation is blocked.

Galactokinase

This enzyme catalyzes the phosphorylation of D-galactose to D-galactose 1-phosphate with the release of ADP, which starts galactose metabolism. There are two galactokinases in humans. The primary enzyme in diverse tissues is GLK1, a monomer of 392 amino acids encoded at 17q24. Galactokinase deficiency with galactosemia, urinary excretion of galactose and galactitol, and cataracts are caused by mutations in its gene. GLK2 is a protein that is encoded on chromosome 15.

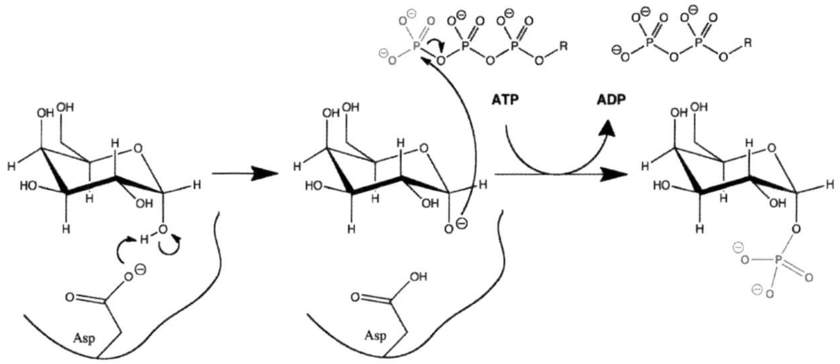

Figure 20. Structure of Galactokinase.

Source: Image by Wikimedia Commons

Galactolipid

Galactolipids are glycolipids that include one or more galactose and/or N-acetylgalactosamine residues. Galactosylceramide and galactosyldiacylglycerol are included in this word.

Galactomannan

Galactomannan is a heteroglycan with D-galactose and Dmannose residues whose structures are still unknown. They're found in the seeds of a variety of plants and trees, most likely as a storage polysaccharide.

Galacse Symbol: Gal.;

The aldohexose galactohexose has two enantiomers, D- and L-galactose. Galactose symbol: Gal; the aldohexose galactohexose has two enantiomers, D- and L-galactose has two enantiomers, D- and L-galactose has two enantiomers, D- and L- D-Galactose (symbol: D-Gal), also known as galactose (brain sugar or cerebrose), is a rare free sugar that is found in numerous oligo- and polysaccharides in plants, animals, and microbes. It is also found in galactolipids and as its glucoside in lactose and melibiose.

Galactose/Glucose Binding Protein

Galactose/glucose binding protein is a bacterial chemotaxis signaling protein.

Galactosemia

Galactosemia, also known as galactosaemia, is a very uncommon illness in humans characterized by a high level of D-galactose in the blood and a unique inability to metabolize galactose. It is caused by a hereditary autosomal recessive trait in which there is a virtual absence of UDPglucose–hexose-1-phosphate uridylyltransferase activity; or, less frequently, by a hereditary autosomal dominant trait in which there is a virtual absence of UDPglucose–hexose-1-phosphate uridylyltransferase activity. A deficit in galactokinase activity is also an inherited autosomal recessive condition.

Galactose Oxidase

Galactose oxidase (EC 1.1.3.9) is a copper-containing enzyme that catalyzes the conversion of D-galactose to D-galactohexodialdose and H_2O_2 via dioxygen oxidation.

Galactosialidosis

Galactosialidosis is a lysosomal storage disorder caused by a lack of three lysosomal enzymes: protective protein/cathepsin A (PPCA), b-galactosidase, and neuraminidase. It is passed down as an autosomal recessive condition. This results in sialooligosaccharide buildup in lysosomes and their excretion in urine.

Galactosidase

Galactosidase can be one of two enzymes: a-galidase (EC 3.2.1.22); otheranother name: melibiase; systematic name: a-D-galactoside galactohydrolase. In a-galactosides, such as galactose oligosaccharides, galactomannans, and galactolipids, an enzyme that hydrolytically eliminates terminal nonreducing a-D-galactose residues.

Galactoside Acetyltransferase

EC 2.3.1.18 galactoside O-acetyltransferase; systematic name: acetyl-CoA: beta- D- galactoside 6-acetyltransferase; other name: thiogalactoside transacetylase; suggested name: galactoside O-acetyltransferase; other name: thiogalactoside transacetylase An enzyme that catalyzes the acetylation of b-D-galactoside to 6-acetyl-b-D-galactoside by acetyl CoA, releasing GDP in the process. The lacA gene product in *E. coli* produces such an enzyme.

B-Galactoside Permease

B-galactoside permease is a term used to describe a group of transport proteins found in bacteria that aid in the absorption of b-galactosides. Melibiose carrier protein from *Klebsiella pneumoniae* is one example. This is responsible for melibiose transport; depending he sugar transported (e.g.e.g., H+-melibiose, Li+-lactose), it can use hydrogen or lithium cations as coupling cations for cotransport, but it cannot recognize Na+ ions despite being an integral membrane protein of the sodium: galactoside symporter family.

Galactosylceramidase

Galactosylceramidase (EC 3.2.1.46) catalyzes the hydrolysis of D-galactosyl-N-acylsphingosine to D-galactose and N-acylsphingosine to D-galactose and N-acylsphingosine. Any D-galayl N-acylsphingosine, i.e.i.e., any cerebroside with D-galactose as the sugar moiety, is referred to as galactosylceramide (less correctly termed galactoceramide or galactocerebroside).

Galactosylceramide Lipidosis

One of a variety of sphingolipidoses or sphingolipid lysosomal storage illnesses is galactosylceramide lipidosis, often known as Krabbe's disease or globoid cell leukodystrophy. Defective activity of galactocerebroside b-galactosidase causes a quickly developing, always fatal genetic (autosomal recessive) illness in human newborns (galactosylceramidase).

Galactosylceramide Sulfotransferase

Galactosylceramide sulfotransferase (EC 2.8.2.11) catalyzes the reaction between a galactosylceramide and 3′-phosphoadenylylsulfate to create galactosylceramide sulphate and adenosine 3′,5′-bisphosphate in sulfatide biosynthesis.

Galactosyl diacylglycerol

Galactosyldiacylglycerol is a kind of lipid in which one or two galactose residues are linked to a 1,2-diacylglycerol by glycosidic linkage. They are found in the thylakoid membrane of chloroplasts and are broadly dispersed throughout plants as component lipids. They are achondria contain them as well.

Galactosyl transferase

Galactosyltransferase is a subclass of hexosyltransferases (EC 2.4.1) enzyme with a galactosyl glycosyl group; the acceptor is usually another carbohydrate molecule or lipid.

Galacturonan

Galacturonan, also known as polygalacturonide, is a glycan made entirely of galacturonic-acid residues; it is a kind of glycuronan.

Galacturonic Acid Symbol: Gala

The uronic acid technically produced from galactose by oxidation of the hydroxymethylene group at C-6 to a carboxyl group is known as galactose uronic acid (GalAGalU or GalUA). D- and L-galacturonic acid are the two enantiomers; D-galacturonic acid is found in plant gums and bacterial cell walls.

Galanin

Galanin is a physiologically active neuropeptide found in a variety of mammalian species' central neurons. The 30-residue sequence is found in human galanin. GWTLNSAGYLLGPHAVGNHRSFSDKNGLTS.

Galaptin

Galaptin is a proteinaceous, beta-galactoside-specific, low-molecular-mass animal lectin that has been isolated from a variety of developing and adult tissues and is thought to facilitate intercellular adhesion in some developmental processes.

Galectin

Galectin is a carbohydrate-binding protein that binds to CD45 and suppresses the tyrosine phosphatase activity of the receptor.

Gall

Gall is an abnormal swelling or growth that a plant produces in response to the activity of numerous fungi, mites, and insects, particularly members of the *Cynipidae* and *Cecidomyidae* families. Galls are frequently discovered to contain insect eggs and larvae.

Gallate

Gallateisaphenolicacidthatistheanionofgallicacid,3,4,5-trihydroxybenzoic acid. 2 any gallic acid salt or ester. In angiosperms, esters and polyesters are found in abundance, mainly as gallotannins.

Gall Bladder

The gall bladder is a hollow, muscular structure that absorbs dilute bile from the liver, concentrates and stores it, and then releases it into the duodenum in many vertebrates.

Gallotannin,

Also known as hydrolyzable tannin, is a tannin made up of a polyhydric alcohol molecule esterified with numerous gallic acid molecules, each of which can be esterified with another gallic acid molecule, and so on.

Gallstone

A gallstone, also known as a cholelith, is a hardened substance that originates in the gall bladder. Light-colored stones, frequently numerous and multidimensional, are the most common kind, composed primarily of cholesterol with some protein and calcium bilirubinate. These are frequently caused by bile stasis, cholesterol supersaturation, and/or the presence of nucleation agents such as glycoprotein in mucus, bacteria, or epithelial cells.

Gamete

A mature reproductive cell whose nucleus (and sometimes cytoplasm) unites with the nucleus of another gamete of identical origin but opposite sex to generate a zygote, which grows into a new individual. Male (+) and female (–) gametes are haploid and distinct.

Gamma Chain

The heavy chain of an IgG immunoglobulin molecule, often know a the gamma chain or c chain.2 Oone of two kinds of subunits in a human foetalhaemoglobin (Hb F) molecule.

Gamma Globulin

A band of serum proteins called gamma globulin or c-globulin migrates the slowest on electrophoresis. In humans, this typically accounts for the greatest proportion of total globulins. Because c-globulins are mostly antibodies and aren't generated in the liver like almost all other blood proteins, the albumin / globulin ratio has long been used to assess liver function.

Galaptin

Galaptin is a proteinaceous, beta-galactoside-specific, low-molecular-mass animal lectin that has been isolated from a variety of developing and adult tissues and is thought to facilitate intercellular adhesion in some developmental processes.

Gamma Helix

A polypeptide chain conformation in which each amide group establishes hydrogen bonds with the group's five residues distant in each direction along the chain is known as a gamma helix or c-helix. It has 5.1 amino acid residues per turn, with 0.99 per residue and 5.03 per turn translation along the helical axis.

Gamma Radiation

During transformation, certain atomic nuclei release gamma radiation or c-radiation, which has wavelengths in the range of 0.1–100 pm (frequencies 3–3000 EHz). Gamma rays have shorter wavelengths and greater photon energy than X-rays. Any abnormal increase in immunoglobulin (typically IgM, IgA, or IgG) or Bence-Jones protein synthesis.

Gammopathies

Gammopathies are characterized by increased immunoglobulin production and can be monoclonal or polyclonal.

Gamone

A gamone is one of a class of biological molecules released by gametes that act on gametes of the opposing sex to start the fertilization process. They might be gynogamones, which are produced by female gametes, or androgamones, whic are produced by male gametes.

Ganciclovir

Ganciclovir, also known as gancyclovir, is a nucleoside analog of acyclovir that is more effective against cytomegalovirus. It is 9-[(1,3-dihydroxy-2-propoxy) methyl]- guanine. It inhibits DNA synthesis by inhibiting DNA polymerase when phosphorylated by virally encoded thymidine kinase.

Ganglion

Outside of the central nervous system, a ganglion is a collection of nerve cells and glial cells.

Ganglioside

Any ceramide oligosaccharide that has one or more sialic residues in addition to other sugar residues is called a ganglioside. Gangliosides were initially discovered in the grey matter of the brain's ganglion cells (neurons), but they are found throughout vertebrate tissues. They're N-acetyl- (or N-glycoloyl-) neuraminosyl-(X) osylceramides, with (X) denoting the neutral oligosaccharide to which the sialosyl residue is connected.

Ganglioside Galactosyltransferase

UDP-galactose–ceramide galactosyltransferase (EC 2.4.1.62); an enzyme that catalyzes the interaction between UDP-galactose and N-acetyl-Dgalactosaminyl-(N-acetylneuraminyl)-D-galactosyl-D-glucosyl-D-glucosyl-Nacylsphingosine to create gangligangliosidosis or a set of hereditary human disorders characterised by ganglioside breakdown abnormalities.

Gap

This is a gap that is added into the written form of a biopolymer sequence of residues to bring a portion of that sequence into register with a comparable piece of sequence in another polymer; it is often employed to accomplish protein or nucleotide sequence alignment.

GAPDH

Glyceraldehyde-3-phosphate dehydrogenase is abbreviated as GAPDH. Because its mRNA is thought to be expressed constitutively in mammalian tissues, it is employed as an internal standard in Northern blotting and PT-PCR.

Gap Gene

The gap gene is one of at least six Drosophila genes that work early in embryogenesis to define the embryo's coarsest subdivisions. Gap genes produce the Krüppel and hunchback proteins, which cause gaps in the segmentation pattern when they are mutated.

Gap Junction

A gap junction or nexus is a specific section of the plasma membranes of apposed vertebrate cells that is 2–4 nm apart and pierced by a connexon that spans the extracellular space and allows free communication between the cytoplasms of the two cells.

Gap Penalty

The cost of adding a gap in a sequence alignment is known as the gap penalty. There may be a single gap penalty or an affine gap penalty with different penalties for beginning and prolonging the gap, depending on the alignment technique utilized.

Gap Phase

During interphase in developing eukaryotic cells, the gap phase is one of two stages, called G1 and G2, in the cell-division cycle during which no DNA synthesis occurs.

Gap Repair

The process of repairing a gap in one strand of a duplex DNA molecule is known as gap repair.

Gas

Regardless of the amount present, a gas is a material whose physical condition is such that it always occupies the whole space in which it is housed.

Gas Chromatography

In gas chromatography, the mobile phase is gas. The word often refers to gas–liquid (partition) chromatography, where the stationary phase is a liquid deposited on a solid support such as a glass or stainless - steel coiled column or the capillary column wall, although it also covers gas–soli chromatography.

Gas Exchange

Ggas exchange between an organism and its surroundings. It involves oxygen intake and CO_2 removal in respiration, as well as CO_2 uptake and oxygen relead gas-filled radiation detector.

Gas–Liqid (Partition) Chromatography

Aa method of separating components of a combination using a stationary liquid phase on a solid support and a mobile gas phase flowing across the liquid phase in controlled way.

Gastric Acid

Tthe hydrochloric acid released by parietal (or oxyntic) cells of the stomach wall.

Gastric Inhibitory Peptide

GIP is a 42-residue polypeptide hormone generated by the K cells of the proximal small intestine in response to glucose in the gut.

Gastric Inhibitory Peptide Receptor

GIP receptors bind gastric inhibitory peptide (GIP) and activate adenylate cyclase. In the rat protein, the signal peptide is residues 1–18, while the receptor is residues 19–455.

Gastric Juice

The acidic liquid released by stomach mucosa glands in reaction to meal intake. Humans have 155 mm of hydrochloric acid in their stomachs, and pepsinogen, the precursor of pepsin proteolytic enzymes.

Gastricsin

EC 3.4.23.3; pepsin C, pig parapepsin II; an enzyme of the pepsin family found in most vertebrate stomach juices. It cleaves Tyr-|- Xaa bonds preferentially. Phosphoprotein has a narrower specificity than pepsin A but considerable activity towards hemoglobin.

Gastrin

A family of linked peptide hormones produced by G cells in the pyloric antral mucosa of the stomach and the proximal duodenum.

Gastrinoma

A pancreatic, stomach, or duodenal G cell tumor. Gastrinomas cause hypersecretion of gastrin and fulminant gastroduodenojejunal ulcerations (Zollinger–Ellison syndrome).

H

The residue of an incompletely specified base in a nucleic acid sequence can be adenine, cytosine, or either uracil or thymine (in DNA).

H1

HNS family of DNA-binding proteins from *Escherichia coli.* H7 1-(5-isoquinolinylsulfonyl)-2-methylpiperazine; an inhibitor of protein kinase C (Ki = 6.0 lM), protein kinase A (Ki = 3.0 lM), and cyclic GMP-dependent protein kinae (Ki = 5.8 lM).

-H

Definition: Aa half-chair structure for a six-membered ring form of a monosaccharide or monosaccharide derivative. Left subscripts indicate locations of ring atoms on the reference plane of the structure, and right subscripts indicate locations on the r side of the reference plane, e.g.e.g., methyl 2,3-anhydro-5-thio-b-LlyxopyranHS.

Habc Domain

Aabbr. HAC domain;domain: a sequence of amino acids in the N-terminal region of syntaxins that forms a triple alpha-helix bundle that folds back on the SNARE domain (which is composed of heptads) to prevent it from forming complexes with other SNAREs. Haber–Weiss reaction is the iron-catalyzed reaction of a superoxide anion with hydrogen peroxide to form oxygen, hydroxyl radical, and a hydroxyl ion.

Haemadin

The Indian leech, *Haemadipsa sylvestris*, produces a 57-amino-acid anticoagulant peptide. Inhibition of thrombin by this slow, tight-binding compound is not observed for trypsin, chymotrypsin, factor Xa, or plasmin.

Hairpin

Any bilayer molecular structure, such as a polynucleotide strand or prostaglandin, which has two adjacent segments folded back upon themselves and is held in that confirmation by secondary molecular forces, such as hydrogen bonds or van der Waals interactions.

Hakai

This protein belongs to the E3 ubiquitin ligase family that specifically binds to the phosphotyrosine residues of E-cadherin to promote its ubiquitination and endocytosis.

Haldane Coefficient

At constant pH, the difference in the number of hydrogens bound by oxy- and deoxyhemoglobin per bound dioxygen molecule.

Haldane Effect

It has been observed that oxygenated blood absorbs less carbon dioxide than deoxygenated blood. It is the reciprocal of the Bohr effect.

Haldane Gas-Analysis

An apparatus once used for the estimation of the carbon dioxide and oxygen contents of a gas sample.

Half-Boat Conformation

Both conformations of a nonplanar monounsaturated six-membered ring compound in which the two ring atoms not directly bound to the doubly bonded atoms lie on the same side of the plane containing the other four (adjacent) atoms.

Half-Cell

It consists of one reversible electrode inserted into a solution. An electrical connection between this solution and that in another half-cell is required to complete the electric circuit.

Half-Chair Conformation

Monounsaturated six-membered rings in either of their conformations when the atoms of the ring that are not directly bound to a doubly bonded atom are on opposite sides of the atoms of the adjacent rings.

Half-Life

T12, or T12; the time in which one-half of atoms of a radilide undergo radioactive decay. AlsoAlso, milar measure of the stability (i.e.i.e., rate of decay) of an excited atom or molecule, a radical, an unstable elementary particle, etc. Additionally, the amount of time it takes to metabolize or excrete one-half of the amount of a drug.

Half-Of-The-Sites Reactivity

Or half-site reactivity is a phenomenon that is shown by many enzymes that exhibit negative cooperativity, in which the maximal stoichiometric yield of either an enzyme-substproduct in a single turnover amountssingle turnover amount to only half the number of apparently equivlent active sites.

Half-Reaction

Eeither of the two coupled chemical changes that together constitute an oxidation-reduction reaction. In one half-reaction there is a gain of electrons and in the other half-reaction a corresponding loss of electrons.

Half-Time Of Exchange

In a reaction involving an exchange, the time is taken for half of the exchangeable atos (or molecules) to be exchanged. Aa half-transporter is a member of the ABC transporter superfamily usually found in peroxisomal or in microbial membranes. It contains a membrane-spanning region (with six transmembrane segments) and a nucleotide-binding domain.

Half-Wave Potential

In polarography, the electrical potential is at the midpoint of the current-voltage curve.

Hallervorden–Spatz Syndrome

A disorder characterized by iron accumulation in the brain, parkinsonism, and dystonia. It is caused by a deficiency of pantothenate kinase.

Halochromism

The accumulation of strongly colored compounds when colorless or faintly colored substances are added to strong acids or the addition of certain metallic salts.

Halogen

Fluorine, chlorine, bromine, iodine, and astatine are monovalent chemical elements in group 17 of the IUPAC periodic table.

Halogenate

The act of treating or reacting (a substance) with a halogen; the introduction of halogen groups into organic compounds.

Haloperoxidase

The enzyme halogenated organic substrates using hydrogen peroxide and a halide ion (Cl-, Br-, or I-). Haloperoxidases contain either heme or vanadium.

Halophile

Organisms that grow in or tolerate saline conditions, particularly in concentrations of sodium chloride equal or greater to that found in seawater.

Halorhodopsina

Identification of a pigment in *Halobacterium halobium* membranes. Upon the absorption of a photon, it catalyzes the inward transport of chloride ions, similar to bacteriorhodopsin.

Halothane

An inhalational anesthetic widely used, volatile, lipid-soluble by name, 2-bromo-2-chloro-1,1,1-trifluoroethane is trivially known as 1-bromo-2-chloro-1,1,1-trifluoroethane.

Hamartin

This protein (1164 amino acids) contains a potential transmembrane segment and a predicted coiled-coil domain but does not have any homology with any other proteins. It may be a tumor suppressor and forms a mainly cytosolic complex with tuberin.

Hamlet

In Drosophila embryos, this protein serves as a genetic switch between single- and multi-dendrite neuron morphology. In its N-terminal half, there are six zinc fingers, and there are three more towards the C-terminus. It has 990 amino acids. It may be a nuclear transcription factor.

Ham's F10 Medium

Based on saline F, Ham's F12 medium formulations for propagating mammalian cells are available.

Handedness

This term refers to objects and compounds which are mirror images of each other and therefore non-superimposable (based on the human left and right hand). A pair of enantiomers displays molecular handedness.

Handle

An additive to any substance to make it more easily identified, reactive, or operable, or as a means to target or bind the substance in question. Chemical or metabolic processes may result in the attachment of a handle.

Handle Technique

In peptide synthesis, excess alkylating agents are separated from the products. By forming a 4-picolyl ester of the carboxyl group at the C terminus of a growing peptide chain, for instance, the C terminus can be protected, resulting in separation of the products on cation-exchange resin columns or extraction into aqueous acidic solutions.

Hanging Drop

Method for crystallizing proteins that involves a droplet of protein solution, bufferipitants being allowed to separateequilibrate by equilibrating with a reservoir of buffers and precipitants.

Hanks' Balanced Salt Solution

HBSS; balanced salt solution in tissue culture that provides the necessary ionic and osmotic conditions for cell growth and development.

Hansch Equation

Relationship between the partitioning differences in an organic and aqueous phase of a compound with the hydrophobicity constant of the substituent.

Haploid

Ploidy of one refers to the number of chromosomes in a cell, organism, or nucleus of a cell which is half the number of chromosomes in a diploid cell.

Hardy–Weinberg Population

In a population consisting of randomly mated individuals, no changes in the frequency of alleles occur from generation to generation.

Hartnup Disorder

The condition is characterized by a specific hyperaminoaciduria and indoluria that results from diminished renal reabsorption of monoamino-monocarboxylic acids, in particular tryptophan. These amino acids may also be ineffectively absorbed through the intestinal tract. It is often associated with a pellagra-like skin rash, which may be resolved by oral nicotinic acid.

Hatch–Slack Pathway

A metabolic pathway for carbon-dioxide translocation in C4 plants, also known as the C4 cycle or C4 pathway. By phosphoenolpyruvate carboxylase, CO_2 dissolves in phosphoenolpyruvate to form oxaloacetate and NADPH is generated to reduce it to malate. The malate passes into the bundle-sheath cells where it is oxidatively decarboxylated by the malic enzyme to pyruvate and CO_2.

Haworth Representation

In diagrammatic form, the structural formulae of cyclic forms of monosaccharides can be illustrated on a plane surface with relative configurations f atom and groups arranged in space.

HBD

Aabbr. fFor a-hydroxybutyrate dehydrogenase, the LD1 isoenzyme of lactate dehydrogenase, otherwise known as 'heart-specific' lactate dehydrogenase, is measured by using a substrate that isn't found in nature.

Head group

Polar lipids express their polarity via their polar chains. For a phospholipid, this comprises the phosphate group together with any polar entity attached to it. A glycolipid includes the carbohydrate molecule.

Headspace Analysis

Microorganisms are identified by gas-chromatographic analysis of the vapor in the headspace above the specimen or culture.

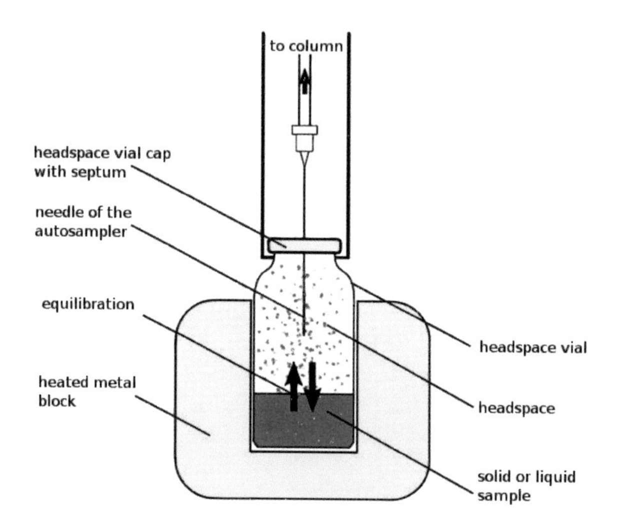

Figure 21. An Illustration of Headspace Gas Chromatography.

Source: Image by Wikimedia Commons

Heat Cpacity Symbol: C

It is usually measured in Jjoules per Kelvin; generally, the amount of heat it takes to raise 1 Kelvin in temperature.

Heat-Labile

Description molecules that decompose at elevated temperatures or that lose biological activity at moderate temperatures, such as 50°C.

Heat Of Activation

A transition state differs from its reactants in terms of heat content.

Heat Of Combustion

When a given amount, usually one mole, oubstance is burned in dioxygen a certain, the amount of heat is evo is known as the hea of combustion. evolves.

Heat Of Formation

Tthe amount of heat absorbed or evolved when one mole of a substance is formed from its elements.

The Heat of Fusion

The heat of fusion or the heat of melting is the amount of heat absorbed or released when a specified quantity of the substance is converted from the solid to the liquid state under given pressure and temperature. If the pressure is 1 atm the temperature is the melting point of the substance

The Heat of Neutralization

The amount of heat absorbed or evolved when one mole of an acid or base is completely neutralized.

Heat of Reaction

Temperature absorbed or evolved during a (bio)chemical reaction, usually expressed as a unit per mole of reactants.

Heat of Solution

As a result of dissolving one mole of a substance in a large volume of specified solvent, heat is absorbed or released.

Heat of Vaporization

Evaporation heat is the amount of energy absorbed or released when a specified amount of a substance, usually one mole, is converted from the liquid to the vapor state at some pressure and temperature. If the pressure is 1 atm the temperature is the boiling point.

Heat-Shock Protein

abbr: HSP is a group of specific proteins synthesized by both prokaryotes and eukaryotes when exposed to high temperatures or other stresses, such as free radical damage.

Heat-Shock Regulatory Element

The heat-shock response element, abbreviated as HSE, is a sequence of bases in prokaryotic and eukaryotic genomes involved in regulating gene expression during heat-shock and related stress situations.

Heat-Stable

When heated to a moderate temperature, e.g. 50°C, something retains its (bio)chemical activity.

Heavy Atom

An isotope, or heavy isotope, or heavy nuclide, contains more neutrons than the most common isotope and is, therefore, larger in mass relative to the most abundant or most common isotope.

Heavy Chain

In immunoglobulin molecules, the H chain is the heavier of the two types of polypeptide chains. Each heavy chain is linked, usually by disulfide bonds, to a light chain and another, identical heavy chain.

Heavy-Chain Disease

Tumors of lymphoid tissue and free immunoglobulin fragments found in plasma and urine are symptoms of this rare human condition.

Heavy-Chain Switch

The change in the expression of the immunoglobulin heavy chain (H chain) gene system manifested by the predominance of IgM (l chain class) in the primary immune response and IgG (c chain class) in the secondary immune response.

Heavy Meromyosin

abbr.: H-meromyosin or HMM; a 350 kDa fragment produced when myosin is subjected to tryptic digestion. It contains part of the helical, rod-shaped tail and both globular heads of the original myosin molecule, and hence retains both the ATPase and the actin-binding activity of myosin.

Heavy Metal

It refers to salts of certain metals with an atomic number greater than 11, usually referring to their toxicity and preference for covalent bonding (hence calcium and magnesium are omitted).

Heavy Nitrogen

The stable nuclide 15 7N (nitrogen-15); relative abundance in natural nitrogen 0.37 atom percent. It is used as a tracer in studies of nitrogen metabolism.

Heavy Oxygen

The stable nuclide 18 8O (oxygen-18), usually in association with the stable nuclide 17 8O (oxygen-17); relative abundances in natural oxygen: 18 8O, 0.20 atom percent; 17 8O, 0.037 atom percent. 18 8O is used as a tracer in studies of reaction mechanisms and oxygen metabolism.

Heavy Strand

Or H strand 1 any polynucleotide chain labeled with a heavy isotope, e.g. nitrogen-15. 2 any naturally occurring chain in a polynucleotide duplex that is heavier or has a greater density than the complementary strand.

Heavy Water Deuterium Oxide

Symbol: 2H2O or D2O; water in which the hydrogen atoms in all of the molecules have been replaced by deuterium; or water in which the hydrogen is appreciably enriched in deuterium.

Hechtian Stand

A threadlike structure that connects protoplasts of plasmolyzed plant cells to their cell wall. Among these are hechtian stands, which are tubes of cytoplasm bounded by plasma membranes, which remain in contact with the cell wall, probably through integrin-type receptors or WAK1 proteins.

Hedgehog Protein

It is a transmembrane protein that is involved in cell-to-cell signaling and segment polarity in *Drosophila melanogaster* embryos and metamorphoses.

HEDTA

abbr. for N-(2-hydroxyethyl)ethylenediamine triacetate, or its trisodium salt, or its partially or fully protonated forms. It is a chelating agent, especially for ferric ions in the pH range 7.0–10.0.

Heisenberg Uncertainty Principle

A particle's velocity (or any related property, such as momentum or energy) cannot be determined with precision simultaneously with its position; the smaller the particle, the greater the degree of uncertainty.

Hela Cell

An established tissue culture strain of human epidermoid carcinoma cells with 70–80 chromosomes per cell (compared to 46 in normal cells). It is much used for biochemical work.

Helicase

A name was sometimes given to the protein (repA) that promotes the ATP-dependent unwinding of the parental DNA duplex during DNA replication.

Helicobacter pylori

The human gastric mucosa is colonized by this Gram-negative organism that is acquired orally as an infant. There are several types and subtypes, with distinct geographical distributions. It is a common cause of gastric ulcers and gastric cancer.

I

Iatrogenic

Describing a condition or disease-induced unintentionally by a physician through his or her diagnosis, manner, or therapy.

IB

abbr. for inhibitors of NF-jB; a cytosolic protein that binds and inhibits NF-jB in most cell types. Upon binding with tumor necrosis factor-a and interleukin-1 receptors, the protein is released from binding by IB kinase complex. The phosphorylation targets IjB for ubiquitination and degradation by proteasomes.

I-Band

abbr. for an isotropic band. The I-bands of striated muscle contain thin filaments and correspond to the light bands. The name derives from the fact that they are isotropic in polarized light.

Ijb Kinase

In the body, NF-jB is released from IjB by a heterotrimeric complex (*700 kDa) and targeted for ubiquitination and degradation by proteasomes. It consists of two catalytic subunits (a and b), which share ≈50% sequence identity, and a regulatory subunit (c).

Ibuprofen 2-(4-Isobutylphenyl)Propionic Acid

Flurbiprofen, ketoprofen, and naproxen are all nonsteroidal antiinflammatory drugs of the substituted propionic acid type.

IC50

It is defined as the concentration that reduces a specified response to 50% of its previous value at the median inhibitory concentration of an antagonist (in mol L–1).

ICAM

abbr. An intercellular adhesion molecule, one of several membrane glycoproteins of the immunoglobulin superfamily. They are ligands for leukocyte adhesion to target cells, in conjunction with LFA-1.

ICE

abbr. An imprint control element consists of a CpG-rich region on genomic DNA that promotes de novo methylation in one parental gamete and prevents it in the other parental gamete. This molecule is necessary for imprinted expression: when unmethylated, it silences mRNA and activates noncoding RNA (ncRNA); when methylated, it is inactive.

ICE-Like Protease

ICE is a family of endopeptidases that structurally resembles interleukin-1b convertase (ICE). The proteins are involved in apoptosis, as they are involved in the proteolysis that causes the death of cells.

I Cell

abbr. Inclusion cells, or CCK cells, refer to cells that produce cholecystokinin and are widely distributed throughout the duodenal and jejunal mucosa. So named because their histological features are intermediate between those of S cells and L cells.

I-Cell Disease

Mucolipidosis II, or inclusion-cell disease, is an autosomal recessive disease in which large inclusions of glycosaminoglycans and glycolipids accumulate within the lysosomes of connective tissue (fibroblasts) due to the absence of certain lysosomal hydrolases.

Ice Point

It is at this temperature that ice melts. It is taken as the temperature (273.15 K) at which ice and water are in equilibrium at standard pressure (101325 Pa). Originally, it was a reference temperature on the Celsius scale, but the Kelvin scale is based on the temperature at the triple point of water (273.16 K).

Ichthyosis

In humans, epidermolysis bullosa is characterized by thickened and scaly skin from a disorder of keratinization in the outer layers of the epidermis. An X chromosome-linked form is caused by mutations that produce a deficiency of steryl-sulfatase.

I-Clip

Intramembrane-cleaving proteases, such as metalloproteases, serine proteases, or aspartate proteases.

Id

abbr. A member of the family of four mammalian helix-loop-helix (HLH) proteins (ID1 to ID4) that act as negative differentiation regulators. They form heterodimers, which do not bind to DNA, with certain ubiquitously expressed basic HLH proteins, and regulate tissue-specific gene expression.

Ideal Gas

Any gas whose behavior is accurately described by the gas laws.

Idiopathic

Idiopathy describes a disease that occurs by itself, not as a result of another disorder, or as a symptom of another disorder; it is an essential disease. Also, idiopathic autonomic neuropathy is a severe subacute disorder caused by autoantibodies to ganglionic acetylcholine receptors.

Idiotype

An antigenic specificity, particularly of antibodies directed against a single antigen.

Idoxuridine 5-Iodo-2′-Deoxyuridine

An analog of the pyrimidine nucleoside that is metabolized to the triphosphate and inhibits DNA replication by substituting for thymidine in viral and mammalian DNA synthesis. It is used as an antiviral agent.

Iduronic Acid Symbol: Idoa

The uronic acid is formally derived from idose by oxidation to a carboxyl group of the hydro oxymethylene group at the position.

IGA Nephropathy

A form of primary glomerulonephritis that is common throughout the world. There is reduced hepatic clearance of IgA, which deposits in the renal glomeruli.

IGA-Specific Serine Endopeptidase

It is the EC 3.4.21.72; another name: IgA protease; the protein that catalyzes the cleavage of immunoglobulin A molecules at the Pro--Xaa bonds of the hinge region. A signal peptide guides the enzyme precursor to the periplasmic space, and the C-terminal helper domain forms a pore in the outer membrane for the exit of the protease domain.

IG Cell

abbr. An intestinal gastrin cell is a small, granular, gastrin-producing cell of the upper small intestine. IG cells are smaller than, and ultrastructurally different from, CCK-producing I cells, gastrin-producing G cells of the gastric pylorus, and TG cells.

Igensin

In mammals, sodium dodecyl sulfate is required for the acceleration of this neutral cytosolic serine proteinase.

Ig-Hepta

A rat glycoprotein (1389 amino acids) is a G-protein-coupled membrane receptor containing two C2-type Ig-homology units with a long N-terminal extracellular region. It occurs on lung cell surfaces and functions in cell adhesion and intracellular signaling. It defines LNB-TM7 as a subgroup of G-protein-coupled receptors.

Ii System

A blood group system, the antigenic determinant is an oligosaccharide related to the ABH system. Anti-I and anti-i are the major cold autoantibodies; the former occurs in the transient anemia accompanying viral or *Mycoplasma pneumonia* infections; the latter is characteristic of the chronic cold agglutinin disease, accompanying chronic lymphoproliferative diseases of the non-Hodgkin type.

Ikaros

Activates the enhancer of a gene CD3-d involved in T-lymphocyte maturation and specification. Similar to the hunchback, it is a zinc-finger protein.

Illegitimate Transcription

In nonspecific cells, a gene that would usually be tissue-specific is transcribing at a low level.

Imidazole 4-Acetate

Imidazole 4-acetate is the byproduct of histamine metabolism; it is formed when monoamine oxidase reacts with histamine to produce imidazole 4-acetaldehyde, which is then oxidized by aldehyde dehydrogenase.

Imidazoline Receptor

An amino or carboxylic acid-binding site in mitochondria or plasma membranes that binds compounds from the generic group's imidazoles, imidazolines, imidazolines, guanidines, and oxazolines endogenous ligand, agmatine. Imidazoline receptors occur throughout the brain, being possibly associated with glial cells, and in peripheral tissues.

Imidazolium

Imidazole cation is formed by adding a proton (hydron) to a molecule of imidazole. The cationic chemical group formed by the addition of a proton (hydron) to an imidazolyl group.

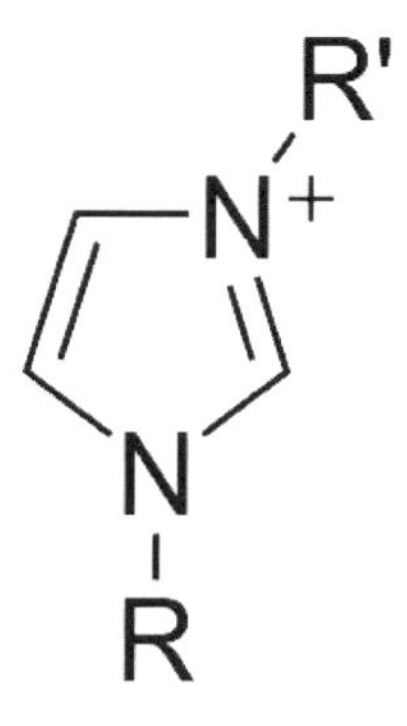

Figure 22. Molecular Structure of Imidazolium.

Source: Image by Wikimedia Commons

Imidazolyl

The chemical group derived formally by removal of one hydrogen atom from imidazole. Typically, hydrogen atoms are removed from the same position on imidazole. The imidazolyl group is present in histidine and its derivatives.

Imide

Any amide that is a secondary organic product (for example, any diacyl derivative of ammonia or a primary amine), especially cyclic compounds where both acyl groups (they may be the same or different) are derived formally or actually from a diacid.

Imidodipeptiduria

An absence of prolidase in the urine is characterized by the presence of peptides containing an amino acid X, commonly glycine. It results from the deficiency of X-Pro dipeptidase and is accompanied by severe ulcers, mostly of the feet and hands, and frequently by mental retardation.

Iminium

In a chemistry compound, a quaternary nitrogen-containing cationic group =+NR2 is attached to carbon, where each R group is hydrogen or any hydrocarbyl group.

Imino Acid

The carboxylic acid in which two hydrogen atoms are substituted with an amino substituent. In biochemistry, the term is commonly applied also to certain cyclic alkylamino (especially a-alkylamino) derivatives of aliphatic carboxylic acids, e.g. proline, although such compounds are now preferably classed as azacycloalkane carboxylic acids.

Iminoglycinuria

The secretion in the urine of excessive amounts of glycine, proline, and hydroxyproline, due to a defect in their shared transport in renal tubules. The condition is clinically benign.

Imipenem

This carbapenem b-lactam antibiotic is a semisynthetic derivative of thienamycin created by Streptomyces cattleya. Its broad-spectrum antibiotic actions resemble those of penicillins, i.e. inhibition of bacterial peptidoglycan synthesis.

Immediate Hypersensitivity

It is the type of hypersensitivity that causes asthma and hayfever as well as some types of eczema. It occurs within minutes of exposure to antigen and depends on activation of mast cells and the release of mediators of acute inflammation. Mast cells bind immunoglobulin E (IgE) at their surface Fc receptors and when an antigen cross-links the IgE, the mast cells degranulate, releasing the substances that are responsible for the symptoms.

Immobilize

In other words, to render any agent, whether it is a particle, a micro- or macrosolute, or an intact cell, non-dispersible in an aqueous medium while maintaining its specific ligating, antigenic, catalytic, or other properties. Immobilization may be achieved by, e.g., encapsulation, entrapment in a small-pore gel, adsorption on, or covalent linkage to, an insoluble supporting material (matrix), or through the formation of aggregates by cross-linkage.

Immortalization

In immortalization, cells with a finite lifespan are transformed into ones possessing an infinite lifespan. It is a characteristic of cancer cells and is of practical importance in, e.g., the creation of monoclonal antibodies.

Immotile Cilia Syndrome

A human genetic disorder is characterized by the immobility or impaired mobility of the cilia in the airways and elsewhere, as well as the tails of sperm. These ciliated cells can be identified by electron microscopy by the absencc of dynein arms in between the microtubular doublets, although other abnormalities can be observed as well.

Immune Adherence

Compounds containing antibodies and antigens, or particles covalently coated with antibodies, to erythrocytes of primate mammals, causing agglutination.

Immune Complex-Mediated Hypersensitivity

Hypersensitivity is caused by the deposition of antigen-antibody complexes in the tissues or blood vessels, also called type III hypersensitivity. The complexes activate, complement and attract polymorphs and macrophages to the site.

Immune Cytolysis

The complement-structured lysis of cells with the aid of using anti-frame molecules. When the cells involved are erythrocytes the methodis called immune hemolysis.

Immune Elimination

Antigens are removed from an immune animal through immune clearance, which involves the complexing of the antigen with antibodies.

Immune Response

The whole immunological response of an animal to an immunogenic stimulus. It consists of antibody formation, the improvement of hypersensitivity, and immunological tolerance.

Immunity Protein

abbr.: Its A protein, synthesized in E. coli with colicin, that binds to the nuclease area of the colicin and inhibits its activity. It is unique for RNase domains to incorporate eighty-four amino acids, which form commonly beta-strands. Those unique for DNase domains incorporate 86 amino acids and shape 4 alpha-helices.

Immunization

Immunization is any process performed on an animal, e.g. the administration of antigen or antibody, that results in an improved reactivity of the animal's immune system in the direction of an antigen or antigens.

Immunoadsorbent

Any insoluble preparation of an antigen (or antibody) appropriate to be used in immunoadsorption.

Immunoadsorption

The use of an insoluble antigen (or antibody) to remove specifically unwanted antibodies (or antigens) from a mixture to make the antibody (or antigen) more specific.

Immunoaffinity Chromatography

A type of affinity chromatography in which one of the two components of an antigen-antibody system is coupled to an insoluble matrix and used for the separation and purification of the other component.

Immunoassay

Any technique for the measurement of specific biochemical substances, commonly at low concentrations and in complex mixtures such as biological fluids, that depends upon the specificity and high affinity shown by suitably prepared and selected antibodies for their complementary antigens.

Immunobiology

The branch of biology dealing with the activities of the cells of the immune system and their relationship with each other and their environment.

Immunobiological

Is a way for the detection of precise polypeptides separated through polyacrylamide gel electrophoresis (PAGE). The bands are transferred from the gel to a nylon or nitrocellulose membrane with the aid of using Western blotting, accompanied by immunological detection of the immobilized antigen.

Immunochemical

Any unique immunological reagent that includes or contains an antigen or an antibody, particularly one of commerce.

Immunochemistry

The department of biochemistry handles the chemical nature of antigens, antibodies, and their interactions, and with the chemical strategies and ideas as implemented in immunology.

Immunochromatography

There are numerous strategies for separating and identifying soluble antigens. In one method, analogous to immunoelectrophoresis, the antigens are first fractionated via a thin-layer chromatogram, e.g. of Sephadex; the fractions are then diffused against a trough containing a solution of antibodies.

Immunocytochemistry

Cytochemistry uses accurately labeled antibody preparations to discover particular cellular components.

Immunodeficiency

Any circumstance wherein there may be a deficiency in the production of humoral and/or cell-mediated immunity.

Immunodiffusion

There are numerous analytical strategies by which components of soluble antigen or antibody combinations can be distinguished. Essentially, the antigens and antibodies are allowed to diffuse in the direction of each other in a translucent gel, in which they react to offer lines or bands of precipitation in characteristic positions.

Immunodominance

The characteristic of a part of an epitope that contributes a disproportionately exceptional portion of the binding energy; e.g. the immunodominance was proven through a monosaccharide residue in determining the antigenic specificity of a polysaccharide

Immunoelectrofocusing

Immune isoelectric focusing is a method that is analogous to immunoelectrophoresis however with the initial fractionation of one of the reactants being affected through electrofocusing in place of electrophoresis.

Immunoelectron Microscopy

A form of electron microscopy wherein structures are stained with particular antibodies labeled with electron-dense material.

Immunoelectropherogram

The document of an immunoelectrophoresis experiment, either the electrophoretic assist itself or a tracing derived therefrom.

Immunoelectrophoresis

abbr.: IE or IEP; a method for isolating and figuring out soluble antigens. It includes region electrophoresis, in a fairly transparent gel, in a single course observed through immunodiffusion against a solution of antibodies (antiserum) located in a trough parallel to the course of electrophoresis.

Immunoenzymatic

About the particular techniques used in immunoenzymology.

Immunoenzymology

The subspecialties of immunochemistry wherein the activity of enzymes coupled to antigens or antibodies are applied as a molecular amplifier of antigen-antibody reactions.

Immunofixation

Immunofixation electrophoresis is a version of immunoelectrophoresis or immuno electrofocusing, wherein proteins of a single immunological species are anchored to the aid through remedy with monospecific antibodies, permitting the alternative proteins to be washed out of the aid.

Immunofluorescence

Immunofluorescence microscopy is a method wherein an antigen or antibody is made fluorescent with the aid of using conjugation to a fluorescent dye after which it is allowed to react with the complementary antibody or antigen in a tissue section or smear.

Immunogelfiltration

A version of radial single immunodiffusion wherein antigen is first fractionated consistent with its molecular size through permeation, in a buffer solution, into a thin layer of the appropriate cross-linked dextran gel. The dextran layer is then protected with a layer of agarose gel containing antibodies. Rings or spots of antigen-antibody precipitate expand in the agarose gel over the complementary antigens.

Immunogen

Any substance that, when introduced into the body, elicits humoral or cell-mediated immunity, but not immunological tolerance.

Immunogenetics

The division of biology which combines immunology and genetics is referred to as immunogenetics. It is the study of genetics that includes the methods and uses of immunological phenomena and substances to future study the use of genetics.

J

Jack-Knife

Jack-knife is a statistical method for evaluating confidence levels of inferred phylogenetic relationships and the difference of assessed evolutionary parameters, in which values are iteratively and arbitrarily eliminated, each new data set is utilized to recalculate relationships.

Jacob–Monod Model

Jacob- Monod model is a model of genetic regulation of protein synthesis in prokaryotes by which the structural genes that decide the organization of the proteins are controlled by different regions of DNA upstream from the structural genes. The latter consists of a control site and a regulator gene, which includes operator (o) and the promotor (p).

JAK

JAK is the abbreviation for Janus kinase. Any of three cytosolic protein tyrosine kinases that bind the intracellular domains of specific membrane receptors. JAK1 is activated by tieing on the non-c polypeptide of receptors for the interleukins IL-2, IL-4, IL-7, IL-9, and IL-15. It phosphorylates STAT proteins other than STAT 5a and 5b. JAK2 is activated on tieing receptors for leptin, somatotropin, and a few cytokines. It then, at that point, phosphorylates STAT proteins, which dimerize and enter the nucleus to activate specific genes.

Jamaican Vomiting Sickness

Jamaican vomiting sickness is a syndrome of poisoning of the nervous system, sickness, and vomiting, associated with metabolic acidosis, profound hypoglycemia, and excessive urinary excretion of glutarate and butyrate, instigated by consumption of the unripe fruit of the ackee, the toxic principle of which is hypoglycin.

Jasmonic Acid

Jasmonic acid is a plant hormone that consists of a C12:1 fatty acid that contains a 6-oxocyclopentane ring formed by C-3 to C-7 and is unsaturated between C-9 and C-10. It is derived metabolically from a-linolenic acid, inhibits growth processes in various tissues, regulates genes involved in the development, and plays a role in plant resistance to disease and insects.

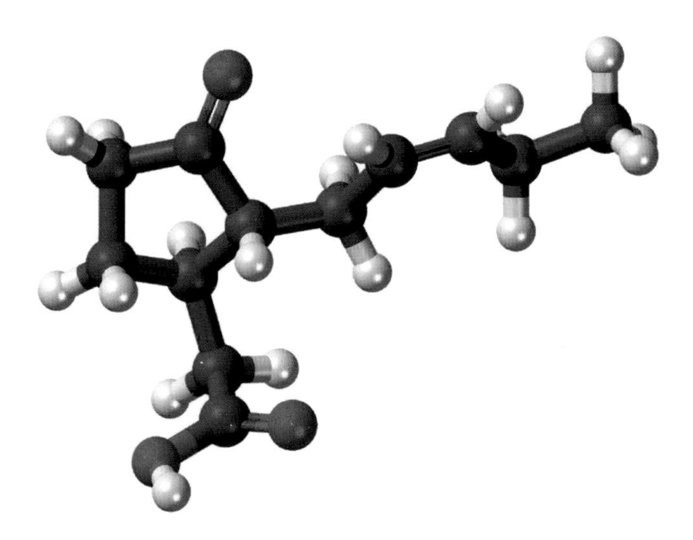

Figure 23. An Illustration of Jasmonic Acid Molecule.

Source: Image by Wikimedia Commons

Jaundice

Jaundice or icterus is a yellowing of the whites of the eyes or/and of the skin indicating an excess of bilirubin in the blood. Jaundice might be classified into three types relying upon the cause: obstructive jaundice in which the passage of bile from the liver to the intestine is blocked. Hepatocellular jaundice, in which there is an inflammation or/and a disease of liver cells thus lessening their capacity to emit bile pigments into the bile passages.

J Chain

J chain or J piece is an abbreviation for joining chain or joining piece. A 15 kDa, cysteine-rich polypeptide occurring in immunoglobulin M (IgM) and immunoglobulin A (IgA) molecules. It cross-links the monomeric immunoglobulin moieties to form a pentamer or dimer in IgM and IgA, respectively. It should not be confused with the J segment of an immunoglobulin gene.

Jekyll

Jekyll is a zebrafish mutant phenotype that indicates defective initiation of heart-valve formation. The mutation includes a homolog of the Drosophila sugarless gene that encodes a UDP-glucose dehydrogenase needed for the production of hyaluronic acid, chondroitin sulfate, and heparan sulfate.

Jelly Roll

A jelly roll is a protein fold comprising of two antiparallel b-sheets placed one above the other to form a sandwich structure. The polypeptide chain follows a route in which successive b-strands alternate between the two sheets. If the b-sheets numbered in parentheses, as (1), (2), etc. and the b-strands are numbered b 1, b 2, etc., the structure of the motif can be represented as loop-b 1 (1)-loop-b 2 (2)-loop-b 3 (1)-loop-b 4 (2)- loop-b 5 (1), and so on.

Jervine

Jervine is a steroidal alkaloid that is found in plants of the genus Veratrum, and it inhibits cholesterol biosynthesis. Pregnant ewes give birth to lambs with congenital malformations if they ingest these plants.

JIP

JIP is an abbreviation for JNK-interacting protein. Another name for this is kinesin-binding protein. Any of three proteins (JIP1 to JIP3) that form part of the JNK signaling pathway and are present in organisms and cells that contain kinesin light chains (KLC). JIP2 and JIP1 are extremely homologous and interact through their C-terminal regions with tetratricosapeptide sequences of KLC.

JM101

JM101 is one of the series of E. coli strains that help the development of bacteriophage M13 vectors. The F′ episome carries lacZM15 making them appropriate for alpha-complementation with plasmids encoding the N-terminal peptide of b-galactosidase.

JNK

JNK is an abbreviation for c-Jun N-terminal kinase. Another name for this is a stress-associated protein kinase which is abbreviated as SAPK. A nuclear protein threonine/serine kinase, belonging to the MAP kinase family, which phosphorylates residues on the N-terminal region of the transcription factor c-Jun. It is actuated by binding to Cdc42 and directs cell development and differentiation, and apoptosis. It is likewise actuated by tumor necrosis factor-a, UV light, and thermal stress.

Johnston–Ogston Effect

Johnston- Ogston effect is a phenomenon happening in sedimentation velocity experiments when at least two solutes are present in a mixture and their sedimentation coefficients are mutually concentration-dependent. It follows that the sedimentation coefficient of the more slowly sedimenting element is less in the plateau region containing the two species than it is in the plateau region between the boundaries, resulting in a higher concentration of the more slowly moving element in the plateau region between the boundaries.

Joining Gene

Joining gene or J segment is abbreviated as J gene. Any of roughly five germ-line genes concerned in coding for immunoglobulin synthesis. The J genes code for the fourth system region of both heavy and light immunoglobulin chains and for some part of the third hypervariable region of the light chain.

Jones–Mote Sensitivity

Jones-Mote sensitivity is a weak and delayed-type hypersensitivity of the skin observed on the challenge a few days after priming with soluble protein in an aqueous solution or in incomplete Freund's adjuvant. The cells infiltrating the lesion characteristically involve a high proportion of basophils.

Jumonji

Jumonji is a gene of humans and mice that is needed for neural-tube formation. The protein contains a C-terminal JmjC domain, an N-terminal JmjN domain, and a dead ringer domain. The two Jmj domains form a functional unit. The JmjC domain also occurs alone in hairless, retinoblastoma protein-binding protein 2, and various putative chromatin-binding proteins.

JUN

JUN is a gene family encoding nuclear transcription factors. V-jun is the oncogene from avian sarcoma virus ASV17 which is named from junana, Japanese for seventeen. JUN encodes transcription factor AP1; it belongs to the fos/jun family and is transiently and rapidly expressed on stimulation of cells by mitogens. The product, Jun, forms heterodimers with Fos, and these, together with Jun homodimers, bind to the AP1 consensus site.

Juvenile hormone

Juvenile hormone or neotenin is an insect hormone, that is produced in the corpora allata glands, and functions to keep the insect in the larval (juvenile) stage. At least three are known, all of them are derivatives of farnesoic acid. They have the potential as insect control agents.

Juxtacrine stimulation

Juxtacrine stimulation is a mode of intercellular communication established by the binding of a membrane-anchored growth factor on one cell to its receptor on an adjacent cell, for example in the case of protransforming growth factor-a. There is no diffusible factor, unlike endocrine, paracrine, or autocrine modes.

K562

An erythromegakaryoblastic leukemia cell line. Cells in culture can be prompted into erythroid advancement by treating them with sodium butyrate.

Kainate receptor

A kind of neuroexcitatory layer receptor described by its particular agonist ligand, kainate (see glutamate receptor), and for which glutamate might be the endogenous ligand. The receptor is coupled to Na+/K+ diverts in the cell film, which are opened when enacted by an agonist.

Kairomone

Any of a heterogeneous group of compound transporters that are transmitted by organisms of one animal group however benefit individuals from another species; they are usually nonadaptive or maladaptive to the producer. They incorporate attractants, phagostimulants, and different substances that intercede the positive reactions of, e.g., hunters to their prey, herbivores to their food plants, and parasites to their hosts

KAL

a human X-chromosome quality that when transformed produces Kallmann condition of hypogonadotropic hypogonadism and anosmia. It encodes a glycoprotein (680 amino acids) of the extracellular grid that is liable for relocation of gonadotropin-releasing chemical (GnRH)- delivering neurons and of olfactory neurons.

Kalata

It is an uterotonic roundabout peptide of 29 amino acids separated from the African plant *Oldenlandia affinis* and a few individuals from the Rubiaceae and Violaceae families. It contains three disulfide bonds and structures just beta-strands.

Kalinin

A fiber related protein of the laminin family, for example (antecedent) human kalinin B1, otherwise called nicein B1 or (most ordinarily) laminin B1k. kalirin a multidomain cytoskeletal protein that capacities as a guanine nucleotide exchange factor. It contains 2959 amino acids, which structure 87 spectrin rehashes, two Dbl and two pleckstrin homology areas, two Ig-like locales, two SH3 districts, and a kinase space.

Kallikrein

Any of two groups of serine endopeptidases that are generally dispersed in mammalian tissues and body liquids, including blood. Kallikrein the particular cleavage of Arg-|-Xaa bonds in little substrates. Its has a specific activity to deliver kallidin.

Kallistatin

Or then again protease inhibitor 4 an acidic protein (in plasma) of the serpin family, communicated generally in human tissues. It restrains tissue kallikrein.

Kanamycin

An aminoglycoside anti-microbial complex from *Streptomyces kanamyceticus*. It is comprised of three parts: kanamycin A (in which R1 = NH2, R2 = OH; see structure), the significant part, frequently assigned as kanamycin; and two minor congeners, kanamycin B (R1 and R2 = NH2) and kanamycin C (R1 = OH, R2 = NH2). Kanamycin causes misreading during protein combination in microscopic organisms, however its objective site on the ribosome is evidently unique in relation to that of streptomycin.

Kanosamine 3-amino-3-deoxy-D-glucose

An amino sugar present in kanamycin and different anti-infection agents, and as free carb in old stocks of *Bacillus aminoglucosidicus*.

Kaolin

Or then again china dirt a finely powdered hydrated aluminum silicate; it is helpful as an adsorbent.

Kaposi's Sarcoma

A skin malignant growth endemic to tropical Africa and frequently happening in AIDS patients.

kappa B

or jB a component normal in eukaryotic promoters. The consensus is GGRNNYYCC, for example, GGGGACTTTCC in mice and humans.

Kappa Chain

or then again j chain image: j; one of the two sorts of the light chain found in human immunoglobulins, the other kind being a lambda chain.

Kappa Convention

An approach to signifying single ligating molecule connections of a polyatomic ligand to a coordination place like a metal particle. Ligands are shown by the italic component image went before by a Greek kappa, j. A right superscript mathematical list demonstrates the quantity of such connections.

Kappa Factor

A protein that causes the end of the release of different bacteriophage DNAs, including those of T4, T5, and T7, by *Escherichia coli* DNA-subordinate RNA polymerase in a DNA site-specific way. It is a dimer of two evidently indistinguishable 17 kDa peptide chains.

Kappa Particle

A cell of any of a few types of Caedibacter present in the cytoplasm of killer strains of *Paramecium aurelia*. At the point when kappa particles (and additionally poisonous constituents or results of them) are delivered by killer paramecia, they might be deadly to sensitive paramecia.

Karplus Curve

A curve portraying the connection of the atomic magnetic reverberation turn coupling steady and the dihedral point.

Karplus Relation

A condition depicting the connection between the extent of the nuclear magnetic resonance turn coupling steady, J, across three bonds and the dihedral point, H, about the focal bond. It is:

$$J = A + B\cos H + C\cos^2 H,$$

where A, B, and C are coefficients that rely upon the electronegativity of the substituents.

Karyophilic Proteins

Soluble proteins that gather in the nucleus on microinjection into the cytoplasm of amphibian oocytes. They are not bound to chromatin or other molecular structures. They incorporate nucleoplasmin and the acidic atomic proteins N1 to N4 of the Xenopus oocyte nucleus.

Karyosome

A compartment, situated on the nucleolus of some plant species, that might uphold the rRNA genes as they stretch out from the nucleolar coordinator area.

Karyotype

The visual appearance of the arrangement of chromosomes of a normal physical eukaryotic cell of a given animal type, individual, or cell strain. It is communicated as far as chromosomal sizes, shapes, and numbers.

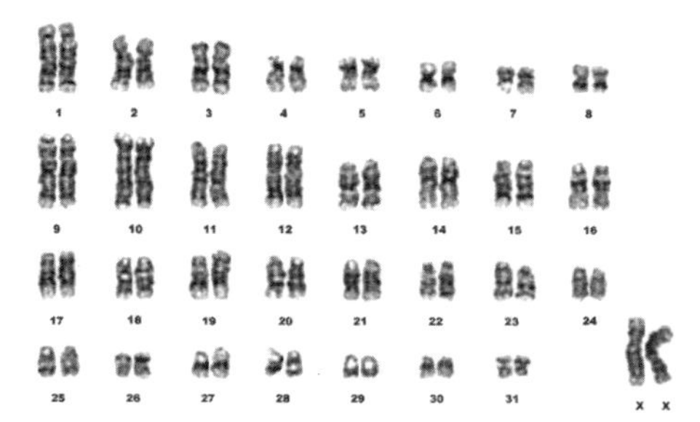

Figure 24. An Illustration of Karyotype of Nine-Banded Armadillo.

Source: Image by Wikimedia Commons

Kasugamycin

An aminoglycoside anti-infection delivered by a strain of *Streptomyces kasugaensis* and containing kasugamine. It associates stoichiometrically with the 30S ribosomal subunit, forestalling commencement of protein biosynthesis in prokaryotes; it additionally hinders interpretation in contagious systems.

Katacalcin

A 21-aminoacid peptide got from the protein antecedent of calcitonin, where it flanks calcitonin on the C-terminal side. It is powerful in bringing down plasma calcium levels. It is available in normal plasma in levels equimolar with calcitonin, and is delivered quickly by calcium implantation.

Katal Symbol

kat; an inferred rational unit for the expression of protein activities, characterized as that synergist movement of a catalyst that will raise the rate of change of a predetermined substance response by one mole each second in a predefined measure system.

Katanin

A microtubule-animated ATPase that cuts off and dismantles microtubules to tubulin dimers.

Kat.f.

An outdated unit for expressing the particular action of catalase preparations, initially characterized as the rate consistent for the catalyzed response under specific circumstances partitioned by the quantity of grams of a compound in the net test volume. The rate steady is determined from the standard condition for a unimolecular response utilizing decadic logarithms and is communicated each moment.

Katharometer

A device used to decide the creation of a gas combination by estimating its heat conductivity.

KATP

abbr. for ATP-subordinate K+ channel; other name: Kir 6.2; a K+ internal rectifier channel of pancreatic B cells that is directed by the ATP-restricting sulfonylurea receptor.

Kauffmann–White Plot

A characterization for the numerous serotypes of the bacterial class Salmonella, characterized as far as their substantial (surface) antigens and, where suitable, their capsular and flagellar antigens.

Kautsky Effect

The trademark changes with time in the fluorescence emission of chlorophyll-a that happen when dark adjusted cells or isolated chloroplasts are illuminated. The progressions might be recognized into a quick change enduring a couple of seconds, and a more slow change enduring a couple of moments.

Kayser

A unit of wave number as the quantity of frequencies of electromagnetic radiation in 1 cm.

Kazal Inhibitor

An electrophoretically heterogeneous proteinaceous trypsin inhibitor and anticoagulant found in the pancreas and in pancreatic juice of various mammalian species.

K Cell

An immunocytochemically recognizable, granule-containing cell, found in the duodenum and jejunum, that produces glucose-subordinate insulinotropic peptide or killer cell a sort of huge granular mononuclear lymphocyte that can kill target cells with an antigen. The K cells draw in the antigen by means of their Fc receptors. They are most likely indistinguishable from regular killer (NK) cells.

K Chromophore

Any chromophore that brings about p-p formation and changes in color.

KCNE

The trait at 21q22.1-q22.2 for a cardiovascular K channel b subunit. It encodes minK, a protein of 129 amino acids, including one transmembrane portion. Various loss-of-function mutations are related to the long QT disorder.

KDEL

A tetrapeptide grouping (in the single-letter amino-acid code) at the C end of proteins that are to be stored in the endoplasmic reticulum (ER). A comparable group is HDEL, particularly in plants. KDEL binds the KDEL receptor, and transcription is believed to be intervened by constant recovery from a post-ER compartment.

KDEL Receptor

A transmembrane protein of the endoplasmic reticulum that binds the KDEL preparation, which comprises the C end of proteins bound to be held in the endoplasmic reticulum. Such receptors are seven-transmembrane-helix proteins. A comparable receptor in yeast for the HDEL group is a result of the quality ERD2.

Kearns–Sayre Syndrome

(abbr.: KSS) or persistent outer ophthalmoplegia (abbr.: CPEO) a mitochondrial infection of people in which ptosis and loss of motion of the outside eye muscles might be joined by myopathy. It is regularly because of a huge deletion (5 kbp) in mitochondrial DNA (mtDNA), and less usually to point transformations in mtDNA for isoleucyl-, asparaginyl-, or leucyltRNA.

Keilin–Hartree Heart-Muscle Preparation

A preparation of submitochondrial particles from cow-like heart muscle that is equipped for catalyzing the reoxidation by dioxygen of NADH and succinate.

Kell System

A blood-group system whose antigenic determinant is essential for an erythrocyte layer glycoprotein of 93 kDa that has an extracellular C-terminal space with 15 cysteine residues, an N-terminal cytoplasmic fragment, and a zinc peptidase site.

Kendall's Compounds

A series of normally occuring steroids of the adrenal cortex and their metabolites; they were temporarily assigned by US physiological scientist Edward Calvin Kendall utilizing letters of the letter set, before the explanation of their functions.

Kennedy's Disease

An elective name for spinal and bulbar solid dystrophy.

Keratan sulfate

Keratan sulfate or (previously) keratosulfate any individual from a group of glycosaminoglycans with basic units comprising of (b1→3)- connected D-galactopyranosyl-(b1→4)- N-acetyl-D-glucosamine and containing variable measures of fucose, sialic corrosive, and mannose units.

Keratein

The dissolvable molecule from the reductive cleavage of disulfide cross-bonds in wool keratin on treatment with sulfhydryl compounds, for example, thioglycolate.

Keratin

Any member from a class of primary strong scleroproteins in vertebrate skin and epidermal designs, for example, feathers, nails, hair, hooves, horns, and plumes, as well as in the cytoskeleton. They are arranged into hard keratins, specific to hair, nails, and quills, and cytokeratins, which structure part of the cell cytoskeleton.

Keratohyalin

A material seen in minute segments in granular cells of the skin. It comprises predominantly of filaggrin.

Kernicterus

An encephalopathy related with degeneration and yellow pigmentation of basal ganglia and other nerve cells in the spinal string and cerebrum, brought about by the serious unconjugated bilirubinemia in babies and once in a while in more youngsters with innate unconjugated hyperbilirubinemia

Ketamine

A dissociative sedative with cataleptic and pain relieving impacts, perhaps through going about as a N-methyl-D-aspartate (excitatory amino corrosive) receptor antagonist.

Ketanserin

3-[2-[4-(4-fluorobenzoyl)- 1-piperidinyl]ethyl]-2,4[1H,3H]-quinazolinedione; a main antagonistto a1 adrenoceptors, H1 receptor receptors, and 5-HT2 5-hydroxytryptamine receptors. A clinically powerful hypotensive specialist, it distinguishes between 5-HT1 and 5-HT2 receptors with high specificity; inside adrenoceptors and receptor receptors it is less specific.

Ketimine

Any imine that is a simple of a ketone; the overall design is R2C=NR′ where R is any organyl group and R′ might be any organyl group or H.

Ketoacidosis

A sort of acidosis brought about by boosted ketone bodies by the ketogenic pathway. This extreme spike brings down blood pH and advances discharge of K+ and Na+ by the kidney, prompting cation exhaustion. It results most frequently from hormonal imbalance, particularly insulin inadequacy, when unsaturated fats are assembled from fat tissue, transported to the liver, and changed over to ketone bodies.

2-Ketoadipic Acidemia

2-ketoadipic acidaemia is an interesting, generally harmless condition where 2-keto, 2-amino, and 2-hydroxy subordinates of adipic acid groups and are discharged in excess levels in urine. These metabolites are gotten from the catabolism of lysine, hydroxylysine, and tryptophan.

Ketoaldonic Acid

Any monosaccharide from an aldonic acid by oxidation of an auxiliary alcoholic hydroxyl group to an oxo group.

Ketoaldose

Any monosaccharide containing both an aldehydic and a ketonic carbonyl group. Ketoconazole is a unique class of imidazole, clinically valuable as an antimycotic. It hinders ergosterol formation in cell films, accordingly expanding their porousness.

Ketodeoxyoctanoate

abbr.: KDO; 2-keto-3-deoxy-D-octanoate; 3-deoxy-D-manno-2-octulosonate; an acid present in lipopolysaccharides of the external films of specific Gram-negative microbes.

Keto-enol tautomerism

Keto-enol tautomerism is a kind of tautomerism in which the keto and enol types of a compound are in balance, the change compared to the movement of a hydron and a shift of holding electrons.

Ketogenesis

The metabolic creation of ketone bodies. The pathway begins with the development of acetoacetyl-CoA (I) from two atoms of acetyl-CoA by acetyl-CoA C-acetyltransferase (EC 2.3.1.9); I then, at that point, responds with a further molecule of acetyl-CoA, in a response catalyzed by hydroxymethylglutaryl-CoA synthase, to form (S)- 3-hydroxy-3-methylglutaryl-CoA (II); II is then cleaved to acetoacetate and acetyl-CoA by hydroxymethylglutaryl-CoA lyase.

Ketogenic Forming

Having the nature of being convertible into ketone bodies by metabolic cycles.

a-Ketoglutarate

A nonpreferred (however frequently utilized) name for 2-oxoglutarate; 2-oxo-1,5-pentanedioate; a compound that plays significant parts in carb and amino-acid digestion, particularly in transamination responses and as a part of the tricarboxylic acid cycle.

Ketoheptose

Any ketose having a chain of seven carbon particles in the atom. Eight enantiomeric sets of such mixtures are possible.

Ketohexose

Any ketose having a chain of six particles in the atom. Four enantiomeric sets of such mixtures are possible, these being the D and L isomers of fructose, psicose, sorbose, and tagatose.

a-Ketoisocaproate

or on the other hand (ideally) a-oxoisocaproate 2-methyl-2-oxopentanoate; a halfway in the biosynthesis of L-leucine. See leucine and valine biosynthesis. a-ketoisovalerate or (ideally) a-oxoisovalerate 3-methyl-2-oxobutanoate; a halfway in the biosynthesis of L-leucine, L-valine, and pantothenic corrosive.

Ketol

Any natural compound that is both a ketone and a liquor, particularly a-ketol (for example acyloin), one in which the two capacities are contiguous.

Ketol Condensation

The catalyst catalyzed reaction of certain ketols (def. 1) by the joining of an acyl moiety of one a-oxo acid to another a-oxo acid to produce a-hydroxy-b-oxo acid. It happens, e.g., in the buildups catalyzed by acetolactate synthase (EC 4.1.3.18); these include either two atoms of pyruvate combining to form a-acetolactate and CO2 , or one particle of pyruvate and one of a-oxobutyrate group to form a-aceto-a-hydroxybutyrate and CO2.

Ketolysis

A term at times used to depict the oxidation, for energy formation, of ketone bodies by extrahepatic tissues.

a-Keto-b-Methylvalerate

a-oxo-b-methylvalerate 3-methyl-2-oxopentanoate; the biosynthetic forerunner of, and catabolic result of, L-isoleucine.

Ketone

Any compound containing a carbonyl group, >C=O, joined to two carbon particles.

Ketone Body

Any of the three substances:

(1) acetoacetate, formed enzymatically from acetyl-CoA;

(2) D-3-hydroxybutyrate (b-hydroxybutyrate), from liver mitochondrial 3-hydroxybutyrate dehydrogenase (EC 1.1.1.30) from acetoacetate; or

(3) CH3)2CO, due to unconstrained decarboxylation of acetoacetate. Ketone bodies might gather in the body in starvation, diabetes mellitus, or affected carb digestion.

Ketonemia

Ketonaemia is a strangely high grouping of ketone bodies in the blood.

Ketonuria

A condition where there is a strangely enormous discharge of ketone bodies in the urine.

Ketopentose

Any ketose having a chain of five carbon atoms in the molecule. Two enantiomeric sets of such mixtures are conceivable, these being the D and L isomers of ribulose and xylulose.

Ketose

aAny monosaccharide of the non-cyclic type of which the true or potential carbonyl group is nonterminal, for example ketonic. The official names and a considerable lot of the inconsequential names of ketoses have the consummation '+ulose'.

L

Label

Label any (radioactive or stable) isotope subbed in an extent of the atoms of a specific compound so the movement or potentially the synthetic or biochemical change of the compound might be followed; marker an effectively conspicuous substance group (which can be radioactive) joined to a particle or consolidated inside a construction.

Labile

It is a synthetic compound that is inclined to unconstrained change or deterioration; somewhat sudden. Or it portrays a nuclear or atomic group in a substance compound that typically promptly separates from the compound; inexactly connected.

Labile phosphate

The sum or extent of any phosphate that is promptly freed from a phosphate-containing compound by hydrolysis in 1 M HCl at 100°C inside 7-10 min. Likewise in some cases known as seven-minute phosphate or ten-minute phosphate. These circumstances were chosen initially for assessing the two terminal phosphate groups of ATP.

Lac

Any of different resinous substances discharged by specific plants or bugs, utilized in polishes.

Lac symbol

For the genetic system of *Escherichia coli* that controls the organism's capacity (lac+) or inability (lac-) to process lactose. See likewise lac operon, lac repressor.

Laccase

Any of a group of copper-protein catalysts (EC 1.10.3.2) of low particularity, that catalyze the oxidation of both o-and p-quinols by dioxygen to form the relating semiquinones and two molecules of water. They are supposed in light of the fact that such a protein was first found in the lac (or veneer) tree (*Rhus verniciflua*); nonetheless, they are currently known to be broadly present in plants and parasites.

Lac I

The underlying lac repressor, the repressor of the lac operon.

Lac I^q

A mutation in the promoterrr of the lacI of *E. coli* that outcomes in increased levels of lac repressor inside cells.

Lac Operon

The lactose operon is a genetic system, or operon, of *Escherichia coli* that controls the formation of the inducible compound bD-galactosidase. It comprises of a straight arrangement (5′ start to 3′ end of the coding strand) of two control positions, called p and o, trailed by three primary genes, assigned z, y, and a; p is the promoter (concerned about RNA-polymerase control), o is the operator (restricting site for the repressor), z is the gene for b-galactosidase, y is the gene for the permease, and a is the gene for thiogalactoside transacetylase.

Lac Repressor

A protein encoded by the lacI gene, that controls the expression of genes coding for chemicals engaged with lactose digestion in the *Escherichia coli* genome; without any inducer, it binds with high specificity to the operator site of the lac operon. It is a homotetramer and specifically binds a 23 basepair DNA in the lac operator through its four N-terminal spaces (headpieces), every one of 51 amino-acid groups.

Lactacin

Any of a group of pore-forming peptidic poisons with around 75 + 65 amino acid deposits, allowing entry of little solutes, including water. They are delivered by *Lactobacillus johnsonii* and are grouped in the TC system under number 1.A.26.

Lactadherin

A Polypeptide medin is a precursor to the protein produced by smooth muscle cells of the aorta. lactalbumin 1 is the fraction of (normally cow's) milk proteins remaining after removal of casein. Specifically, it contains beta-lactoglobulins and a-lactalbumins. 2 a-lactalbumin is a Ca2+- binding protein that occurs as a minor constituent of milk of many mammals; its 123 amino-acid sequence is highly similar to that of lysozyme.

Lactam

In another amino acid than the a-amino acid, any intramolecular cyclic amide is produced by formally removing a molecule of water between the amino and carboxyl groups. A prefix (as in b-lactam) may be habituated to designate the position of the amino group in the parent compound. A lactam commonly subsists in tautomeric equilibrium with its corresponding lactim.

ß-Lactam

The antibiotics of a particular class are those with the characteristic feature of a four-membered b-lactam ring. They include the penicillins and many of the cephalosporins.

b-Lactamase

b-lactamhydrolases are constitutive or induced enzymes (EC 3.5.2.6); systematic name: b-lactamhydrolases. They are coded by certain strains of many bacterial species and can catalyze the hydrolysis of the lactam linkage in one or more b-lactam antibiotics, leaving a superseded b-amino acid; they thereby inactivate the antibiotic.

Lactam

During lactam-lactim interconversion, lactim tautomerism occurs as a result of the migration of a hydrogen atom between the nitrogen atom and the oxygen atom. It represents a special case of amide–imidol tautomerism.

Lactase

A common name for any b-D-galactosidase. Some lactases hydrolyze other glycosides, including a-L-arabinosides, b-D-glucosides, and b-D-glucosides.

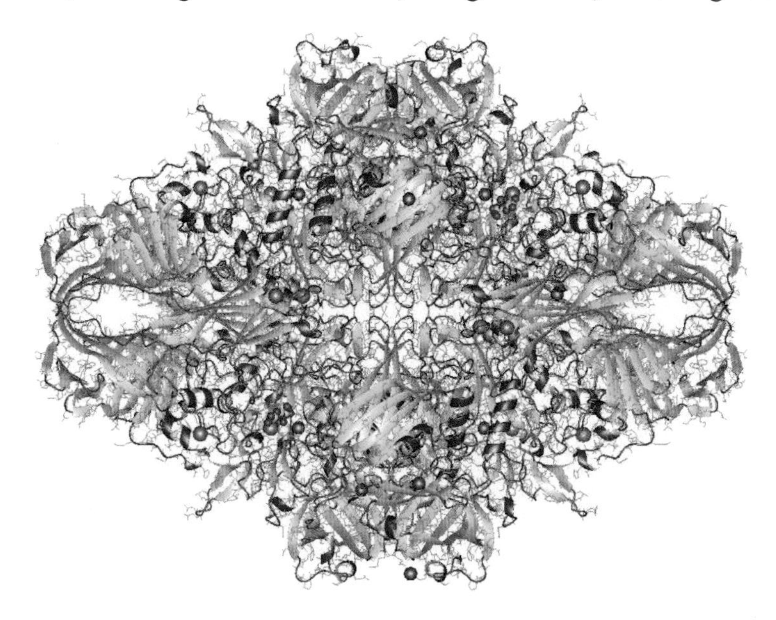

Figure 25. Structure of Lactase.

Source: Image by Wikimedia Commons

Lactase Deficiency

This is a congenital deficiency that occurs in humans when there is a deficiency of lactase (b-galactosidase) in the small intestine, resulting in the inability to digest lactose and intolerance to it. People of all ethnicities

(45–95%) are affected, but those of central and northern European ancestry (*25%) are more likely to experience the condition. The prevalence of the condition varies considerably, but it is more prevalent in native Australian and Oceanic populations, east and south-east Asian populations, tropical African populations, and Americans compared to those of other ethnicities.

Lactate

a-hydroxypropionate; 2-hydroxypropanoate; the anion, CH_3– CH (OH) – COO^-, derived from lactic acid. It occurs naturally as D(–)-lactate (i.e. (R) – 2 – hydroxypropanoate) and L(+) -lactate (i.e. (S) – 2 – hydroxypropanoate). L(+)-lactate is formed by anaerobic glycolysis in animal tissues, and DL-lactate is found in sour milk, molasses, and certain fruit juices.

Lactate Dehydrogenase

Lactate and pyruvate are converted by an enzyme. L-lactate dehydrogenase (EC 1.1.1.27; abbr.: LD or LDH) acts on L-lactate and NAD+. It occurs as homotetramers or heterotetramers of the heart muscle (H) and skeletal muscle (M) forms (the five isoenzymes are: LD1, H4; LD2, H3M; LD3, H2M2; LD4, HM3; and LD5, M4).

Lactation

Human milk is secreted through the mammary glands. It is a consummate period of milk secretion from about the time of parturition to that of weaning.

Lacteal

1. of, relating to, or resembling milk; milky.
2. (of lymph vessels) containing or transporting chyle.
3. any blind-ended lymph vessel that carries chyle from a villus of the small intestine to the thoracic duct.

Lactic

Originating from or relating to milk. lactic acid a-hydroxy propionic acid; 2-hydroxypropanoic acid; a reasonably strong acid (pKa = 3.86 at 25°C). Lactate is most commonly seen in biological systems as its anion. It is also useful as a preservative of foodstuffs.

Lactic-Acid Bacterium

Any main metabolic product where lactate is the main bacterium. Such bacteria belong chiefly to the genera Lactobacillus and Streptococcus, and some are of commercial importance, e.g. in the manufacture of fermented foods.

Lactic-Acid Fermentation

Heterolactic fermentation and/or homolactic fermentation is a more common name for lactic acid fermentation. A common cause of lactic acidosis is persistently raised blood lactate concentrations, usually above 5 mmol L–1. Other factors that contribute to lactic acidosis are vigorous exercise, poor cardiovascular function, hypoxia, shock, and some inborn errors of metabolism.

Lactic

Lactic, in general, refers to or derived from milk. Lactic acid is a 2-hydroxypropanoic acid, which is a reasonably strong acid having pKa = 3.86 at 25°C. In the biological systems, it mainly exists as its anion 'lactate'. In addition, it is used as a preservative of food items.

Lactide

Two (or more) molecules of a hydroxycarboxylic acid of any intermolecular cyclic ester are composed by the process of self-esterification.

Lactifer

Rubber latex accumulates in a special cell, found in the bark of the rubber tree (Hevea).

Lactim

It is any intramolecular compound derived from the removal of a molecule of water between an amino group and a carboxyl group of an amino carboxylic acid other than an amino acid. A prefix (as in b-lactim) may be habituated to designate the position of the amino group in the parent compound. A lactim commonly subsists in tautomeric equilibrium with its corresponding lactam.

Lactitol

4-O-b-D-galactopyranosyl-D-glucitol; a disaccharide used as a sweetener.

LactobacillicAcid

(1R-cis)-2-hexylcyclopropanedecanoic acid; 11,12-methyleneoctadecanoic acid; a lipid constituent of a large number of microorganisms, including *Lactobacillus arabinose.*

Lactobacillus

A genus of Gram-positive, anaerobic, or facultatively aerobic, straight or curved bacilli or coccobacilli. all species form lactate as the major end product of glucose fermentation which is present in their metabolism. certain fermented foods, e.g. cheese, yogurt, sauerkraut contain various species of lactobacilli which helps with the preparation

Lactobionic Acid

The aldonic acid is obtained by the oxidation of lactose.

Lactochrome

Also known as riboflavin. There are about 25 amino acids residues in lactocin, a pore-forming peptidic toxin that permits the passage of minuscule solutes, including dihydrogen monoxide. They are engendered by Lactobacillus sake and are relegated in the TC system under number 1.A.23. lactococcin designation of three different families of peptidic pore-forming toxins, engendered by Gram-positive bacteria.

Lactoferrin

Lactoferrin is transferrin present in milk. Its level in human plasma may be elevated in pancreatitis.

Lactonase

recommended name: gluconolactonase; systematic name: D-glucono-1,5-lactone lactonohydrolase; other name: Aldo lactonase. An enzyme that catalyzes the reaction: D-glucono-1,5-lactone + H_2O = D-gluconate;

Lactone

The chemical procedure of removing a molecule of water from between the two hydroxyl and carboxyl groups of a hydroxycarboxylic acid other than an a-hydroxy acid to produce an intramolecular cyclic ester. A prefix (as in b-lactone) may be used to designate the position of the hydroxyl group in the parent compound.

Lactoperoxidase

A special name for peroxidase from milk. It withal catalyzes the oxidation of iodide to iodine, hence it is utilizable in radioiodine labeling of proteins and other biological materials under mild conditions.

Lactose

Milk sugar is the nugatory name for the disaccharide 4-O-b-D-galactopyranosyl-D-glucose. It constitutes roughly 5% of the milk of virtually all mammals (human milk 6.7%; cow's milk 4.5%). It occurs infrequently in plants, e.g. in the anthers of *Forsythia* spp.

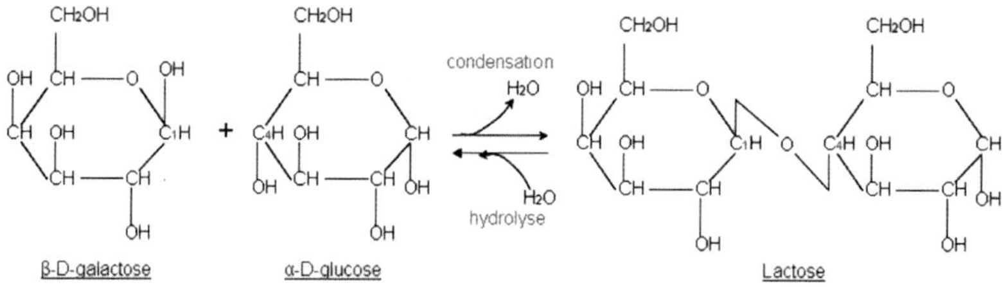

Figure 26. Formation of Lactose.

Source: Image by Wikimedia Commons

Lactose Fermenter

Any genus, species, or strain of microorganism that ferments lactose is termed as jargon.

Lactose Intolerance

A syndrome occurring in humans with congenital or acquired intestinal lactase deficiency with the ingestion of lactose (in milk). The failure to hydrolyze lactose leads to its accumulation in the intestine and consequent osmotically induced net fluid secretion into the gut, expeditious intestinal transport, and acid stools due to bacterial fermentation of lactose. The condition is concretely prevalent in the Orient.

Lactose Permease

An H+–lactose symporter in *Escherichia coli.* It has 12 transmembrane helices but is not very proximately cognate to most other members of the superfamily.

Lactose Synthase

EC 2.4.1.22; systematic name UDP galactose: D-glucose 4-b-D-galactotransferase; another name: UDP galactose–glucose galactosyltransferase. An enzyme that catalyzes the reaction: UDP galactose + D-glucose = UDP + lactose. In the mammary gland, the protein comprises N-acetyl lactos amine synthase, the A protein, and a-lactalbumin (visually perceive lactalbumin), the B protein, both of which are required for lactose synthesis.

Lactose-Tolerance Test

A procedure utilized in the investigation of suspected lactase deficiency. Blood-glucose concentrations are estimated before an oral dose of lactose (1 g per kg body mass) and thereafter at 30-min intervals for 2 h. In mundane individuals, an incrementation of at least 30 mg per 100 ml (0.167 mM) in the blood-glucose concentration above the fasting value is visually examined.

Lactosuria

The presence of lactose in the urine. It is not unconventional during late pregnancy and lactation, or if the flow of milk is prevented.

Lactosyl

The glycosyl groups formed from lactose by detaching the anomeric (a or b) hydroxyl group on C-1.

Lactotroph

A class of irregular, granulated, acidophilic cells of the anterior pituitary gland that secrete prolactin. The number increases greatly during pregnancy.

Lactotrophic

Any agent that acts on the mammary gland to induce changes in its metabolism.

Lactoyl

The acyl group, CH3–CH(OH)–CO–, is derived from lactic acid.

Lactulose

Lactulose the common name for 4-O-b-D-galactopyranosyl-D-fructofuranose; is a semisynthetic disaccharide, prepared from lactose, that is sweeter than lactose but less saccharine than sucrose. Concentrated solutions of lactulose are used clinically as an osmotic laxative. LacY is found in a gene for lactose permease in *Escherichia coli.*

LacZ

A gene in the lac operon of *E. coli* encoding β-galactosidase, it is widely used as a reporter gene.

lacZdelta M15

A mutant form of the E. coli β-galactosidase gene in which the sequence encoding the N-terminal region corresponding with the α-peptide is deleted. *E. coli* strains carrying this deletion are used to test for a-complementation.

Ladder informal

The 1kb ladder is an example of one of many types of DNA molecular size markers.

L1

Adhesion molecule is a type of I membrane glycoprotein involved in neuron–neuron adhesion, neurite fasciculation, the outgrowth of neurites, etc. It is a member of the immunoglobulin superfamily and has fibronectin domains.

Laforin

Many tissues contain protein-tyrosine phosphatase. Mutations lead to Lafora disease, progressive myoclonus epilepsy with astute decline, accumulation of glycogen in encephalon, skin, and liver, and early death.

Lagging strand

The DNA strand that is synthesized by the joining together of short Okazaki fragments, giving the effect of perpetual 3′ to 5′ synthesis in the direction of fork during the discontinuous replication of duplex DNA.

Lag Phase

The initial phase of the multiplication of a bacterial culture during which the rate of incrementation of cell numbers remains static, before ascending to a value determined by environmental conditions. The extent of the lag phase is determined by the cultural history of the cells of the inoculum and the chemical and physical conditions of the medium into which they are placed.

Lambda Chain

Also known as the k chain symbol: k; one of the two types of the light chain of human immunoglobulins, the other type being a kappa chain.

Lambda DASH

It is used for cloning DNA which is a replacement of a proprietary bacteriophage k vector that fragments up to 20 kb.

Lambda FIX

A proprietary bacteriophage k replacement vector used for cloning DNA fragments of up to 20 kb. lambda gt10 a bacteriophage k insertion vector used for cloning DNA fragments of up to 8 kb.

Lambda gt11

A bacteriophage k insertion vector, designed to allow expression of proteins encoded by DNA inserts .it is also used for cloning DNA fragments of up to 8 kb.

Lambda Insertion Vector

One of several bacteriophage k cloning vectors in which the cloned sequence is inserted into a single restriction endonuclease site. The size constraints on packaging bacteriophage k recombinant DNA limits the size of the insert.

Lambda Phage

It is a temperate phage that infects *Escherichia coli* and contains linear duplex DNA, which circularizes during infection; the genome is 48 514 bp. Consequential gene clusters include those for the head and tail genes, recombination (att, int, xis, a, b, c), lysis, and regulatory genes.

Lambda Pipette

A pipette used for the transfer of volumes in the microlitre range.

Lambda Replacement Vector

This is among several bacteriophage k cloning vectors designed for cloning DNA inserts of up to 20kb. The size constraints for packaging bacteriophage k recombinant DNA dictate the need to remove an extra fragment at the junction of two homologous restriction endonucleases before insertion of the DNA to be cloned. They are used for cloning fragments of genomic DNA.

M9

A combination of inorganic salts solution used as the basis of minimal growth media for bacteria.

MAC

abbr. for 1 membrane attack complex. Monocytes, macrophages, and granulocytes express MAC 1, a cell-surface glycoprotein that participates in adhesion interactions and plays a role in complement-induced uptake by these cells. It probably recognizes the RGD peptide in complement component C3b. It is a heterodimer of aM and b2 chains.

Mackerel

The subfamily Scombroidei of any marine fish. These fish are rich in oils with a relatively high content of n-3 fatty acids. They are as significant as any food fish.

Mackerel

Any marine fish of the subfamily Scombroidei. These fish are of importance as food fish, and their flesh is rich in oils with a relatively high content of n-3 fatty acids

Macrocyclic

Having a long - life cycle. macrocytic anemia in which the circulating erythrocytes are more larger than normal ones. It is often due to a deficiency of vitamin B12 (cobalamin) or folic acid.

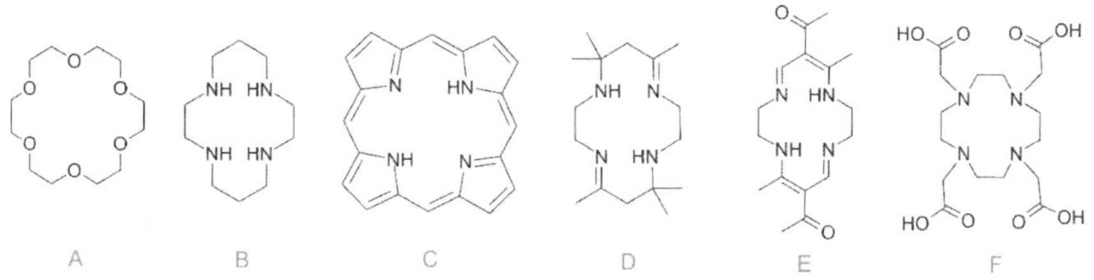

Figure 27. Different Macrocyclic Structures.

Source: Image by Wikimedia Commons

Macroglobulin

Any plasma globulin of >400 kDa. Macroglobulins include immunoglobulin M, a2 -macroglobulin, and many lipoproteins. a2-macroglobulin abbr.: a2 -M; a glycoprotein found in vertebrate plasma that strongly inhibits proteases of all classes. In humans, it constitutes about one-third of the a2 -globulins; its synthesis is enhanced in hypoalbuminemia when it constitutes a major proportion of the increased a2 -globulin band on electrophoresis bands.

Macroglobulinemia

A condition in humans called macroglobulinemia in which there is an increase in the concentration of macroglobulins, especially immunoglobulin M, in the blood. It is often associated with lymphocyte tumors.

Macro Histone 2A

abbr.: macro H2Aa; the obverse of histone H2A that supersedes this in certain nucleosomes, with an implicatively insinuated role in nucleosome situating. macroion is an ion of a macromolecule, especially one that does not pass through a dialysis membrane.

Macrolide

Any antibiotic produced by *Streptomyces* spp. that contains a large lactone ring (see +olide) with few or no double bonds and no nitrogen atoms, linked to one or more sugar moieties. The macrolides include the carbomycins, the erythromycins, oleandomycin, oligomycins, and spiramycins. They are active mainly against Gram-positive bacteria and they inhibit the early stages of protein synthesis.

Macrometabolite

A system with a reasonably high concentration of any metabolite, especially as opposed to a trace metabolite, i.e. one present in extremely small concentrations.

Macromolecule

The term describes molecules with an extremely large number of atoms, operationally defined as molecules with masses greater than 10 kDa (and up to 102 MDa or more) that cannot pass through the pores of dialysis tubing as it is generally used. The term includes nucleic acids and most polysaccharides and proteins.

Macronutrient

Any element that is required by living organisms in relatively immensely colossal quantities for normal magnification.

Macrophage

Any cell of the mononuclear phagocyte system that is characterized by its facility to phagocytose peregrine particulate and colloidal material. Macrophages occur in connective tissue, liver (Kupffer cells), lung, spleen, lymph nodes, and other tissues. They contain prominent lysosomes, stain with vital dyes, and play a consequential part in nonspecific immune reactions.

Macrophage Inflammatory Protein

abbr.:MIP; a cytokine engendered by B and T cells, monocytes, mast cells, and fibroblasts. MIP 1a is a chemoattractant for monocytes, T cells, and eosinophils; it withal inhibits hematopoietic stem cell engendering. MIP 1b is a chemoattractant for monocytes and T cells and is an adhesion molecule, binding to b1 (VLA family) integrins.

Macrophage Migration Inhibitory Factor

abbr.: MIF; another name: delayed early response protein 6; a lymphokine that is produced by primed lymphocytes on incubation with the priming antigen. It inhibits the migration of macrophages out of macrophage-rich tissue such as the spleen and is probably a mediator of macrophage participation in inflammation.

Macrophage Scavenger Receptor

The membrane receptor for oxidized low-density lipoproteins (LDL) on macrophages and sinusoidal endothelial cells. It is implicated in the deposition of cholesterol into arterial walls during atherogenesis. Withal is a membrane receptor of hepatocytes and steroid hormone-engendering cells that binds high-density lipoproteins (HDL), selectively abstracts cholesterol and mediates its uptake.

Macro Pipette

A pipette of relatable construction to a standard plunger-type micropipette but of capacity in the range of 0.25–5 mL: macropore any pore (in a gel or porous solid) whose width is more preponderant than about 50 nm.

Micropore

Any pore, either in a gel or porous solid whose width does not exceed around 2.0 nm.

Macroscopic Equilibrium Constant

Any constant describing the overall equilibrium between two chemical entities that may be interconverted through two or more alternative intermediates.

Macrosolute

Describes a substance larger than the size at which it can pass through a membrane with an established pore size or permeability limit.

Macrotetrolide

Or macrotetralide is any antibiotic that contains a 32- membered ring built up of four hydroxy-acid residues and containing four ether and four ester bonds; the group includes nonactin and related antibiotics. Some similar compounds have an open-chain structure, e.g.monensin and nigericin.

MAD 1

A basic helix-loop-helix transcription factor capable of forming dimers with MYC and MAX. 2 abbr. for multi-wavelength anomalous dispersion.

MAD2

A protein in yeast that is an essential component of the spindle checkpoint during mitosis and determines the fidelity of chromosome distribution. Mutation leads to a large increase in the rate of chromosome misaggregation during meiosis.

MadCAM

abbr. for mucosal addressin cell adhesion molecule; MadCAM 1, Mr 58–66 kDa, is a member of the immunoglobulin superfamily of adhesion molecules. It is constitutively expressed on endothelial cells of post-capillary venules. It mediates lymphocyte homing to mucosal lymphoid organs and lamina propria venules, and binds L-selectin and integrin a4b7.

Mad Cow Disease

An informal name for bovine spongiform encephalopathy.

Madder

The ground root of *Rubia tinctoria*, used as a plant dyestuff; the principal component is ruberythric acid.

MADS box

A highly conserved amino acid sequence (≈56 amino acids) present in the floral 'homeotic' transcription factors APETALA1, APETALA3, PISTILLATA, and AGAMOUS of *Arabidopsis thaliana* and in many other proteins expressed in flowers of other plant species.

Magainin

Any antibiotic peptide from Xenopus skin whose pore-forming activity (see PFP) results in the permeabilization of bacterial membranes. Magainins have a wide spectrum of action against bacteria, protozoa, and fungi. Magainin I has the sequence GIGKFLHSAGKFGKAFVGEIMKS; magainin II is (Lys10, Asn22)-magainin I. An analog with greater antibiotic potency is (Ala8,13,18)-magainin II amide. Magainins are classified in the TC system under number 1.C.16.

Magnesium

Magnesium symbol: Mg; a metallic element – an alkaline earth metal – of group 2 of the IUPAC periodic table; atomic number 12; relative atomic mass 24.305. It occurs naturally only in the combined state and is a mixture of stable nuclides of mass 24, 25, and 26. Magnesium is one of the most abundant elements of the Earth's crust (2.1%) and an essential component of all living material. It forms a divalent cation, Mg2+, which is a component of chlorophyll, occurs in bone, and is an essential cofactor for many enzymes, including the majority of enzymes utilizing ATP. It is a major biological cation, the fourth most abundant cation in the human, much of it being in the skeleton; the range in normal human plasma is 0.8–1.2 mmol L–1. An excretion mechanism in the kidney is partly regulated by parathyroid hormone.

Figure 28. An Illustration of Magnesium.

Source: Image by Wikimedia Commons

Magnetic Beads

They are microscopic synthetic polymer beads incorporating a magnetite core and chemically modified to enable the covalent attachment of proteins such as antibodies. Beads coated with streptavidin will bind proteins or DNA labeled with biotin. Such beads can be concentrated from suspension by exposing them to a magnetic field thereby providing an alternative to centrifugation.

Magnetic Circular Dichroism Spectroscopy

Magnetic circular dichroism spectroscopy measurement of circular dichroism of a colored material which is induced by a magnetic field applied parallel to the direction of the measuring light beam. Even achiral materials exhibit MCD (the Faraday effect). Variable-temperature MCD is used for identifying and assigning electronic transitions originating from paramagnetic chromophores.

Magnetic Resonance Imaging

abbr.: MRI; an imaging technique that uses nuclear magnetic resonance (NMR) to investigate the state of tissues in the intact body; it is so-called to avoid the use of the term 'nuclear' in the treatment of patients. It has the advantage that it is non-invasive. A magnet large enough to surround a part of the body is used in conjunction with computer analysis to provide an image resulting from the NMR signals. The technique can distinguish extracellular water from that inside cell on the basis of protein spin relaxation rates; this distinction is used to reveal certain types of tissue damage.

Magnetic Shielding

Protection from the effects of an external magnetic field. Iin nuclear magnetic resonance spectroscopy, the reduction of the applied magnetic field at an atomic nucleus (as compared with that applied to the whole sample) brought about by the electron cloud of the molecule.

Magnetic Stirrer

Any apparatus for stirring a liquid in which a follower – usually a small bar magnet coated in glass or plastic – may be caused to oscillate or rotate within the liquid under the influence of a more powerful magnet outside the container. Motion of the external magnet is controlled by a variable-speed electric motor. The term is sometimes applied specifically to the follower

Magnification

The apparent linear enlargement of an object when viewed through a lens, system of lenses (as in a microscope, telescope, etc.), or other instruments. It is given by the ratio of the apparent diameter of the object, as seen using the lens, to its real diameter, as seen unaided.

Maillard Reaction

Maillard reaction is a nonenzymic reaction in which a reducing sugar combines at its anomeric carbon atom with an amino group of an amino acid, peptide, or protein to form first a Schiff base and then, by an Amadori rearrangement, the corresponding ketoamine. It is the first of a series of reactions that may occur during the processing and/or storage of food, resulting in a loss of nutritive value and the formation of a brown coloration; hence it is also known as the browning reaction. It may also be the reaction leading to the formation of hemoglobin A1c in the erythrocyte.

Main Protease

It is the protease that cleaves the polyprotein encoded in the RNA genome of coronaviruses. The human virus's main protease (305 amino acids) contains two six-stranded beta barrels and a cluster of five alpha-helices. It cleaves on the carboxyl side of glutamine at about a dozen sites that contain Leu–Gln–Ala/Ser/Gly sequences.

Major Facilitator Superfamily

Major facilitator superfamily abbr. MFS, one of the two largest families of membrane transporters (the other being the ABC transporters). Transporters of this family are present ubiquitously in bacteria, archaea, and eukarya. They transport a wide range of metabolites and drugs, including sugars, amino acids, nucleosides, and a large variety of anions and cations. All permeases of the MFS possess 12–14 putative transmembrane a-helical regions, and exhibit a well-conserved motif present between transmembrane regions 2 and 3.

Major Histocompatibility Complex

Major histocompatibility complex abbr.: MHC; a complex of genetic loci occurring in higher vertebrates that encodes a family of cellular antigens, known in mice as histocompatibility-2 antigens and in humans as human leukocyte-associated antigens. The MHC antigens are cell-surface glycoproteins and may be divided into two classes, designated I and II. In transplantation reactions, cytotoxic T lymphocytes respond mainly against foreign MHC glycoproteins of class I together with antigen, while the response against class II and antigen is mainly by helper T lymphocytes

Makefile

Makefile a file read by the UNIX utility make. It includes instructions for the computer and is used largely, but not exclusively, for compiling programs from source code. The advantage of this is that instructions on files that have not been altered since the last use of make are not repeated.

Malate–Aspartate Cycle

Malate–aspartate cycle or malate cycle an intracellular metabolic cycle, also known as the malate–aspartate shuttle or malate shuttle, that can transfer reducing equivalents from the cytosol to the mitochondria in liver and other tissues. Oxaloacetate formed inside the mitochondria is there converted by a transaminase into aspartate, which passes through the mitochondrial membrane on a specific carrier; a similar transaminase in the cytosol converts this aspartate back into oxaloacetate, which then reacts with cytosolic NADH, in a reaction catalyzed by cytosolic malate dehydrogenase, to form (S)- malate and NAD+. The malate thereupon passes back into the mitochondria, where it is dehydrogenated again to oxaloacetate, with concomitant reduction of mitochondrial NAD+ to NADH and completion of the cycle

Malignant Hyperthermia

Malignant hyperthermia is a genetic disorder of humans and pigs that manifests as a fulminant episode of muscle rigidity, high fever, and acidosis, which is frequently fatal. In humans, there is an autosomal dominant predisposition, with general anesthesia (frequently involving halothane inhalation) as the trigger. Most cases are associated with hyperexcitability of the ryanodine receptor. In pigs, the predisposition is autosomal recessive with stress as the trigger, and is associated with a point mutation in the same receptor.

Maleylation

It is the introduction of one or more maleyl groups into a substance by acylation, e.g. with maleic anhydride. Maleylation of a protein with maleic anhydride is used to acylate its free lysine (and other) amino groups in order to change their charge at neutral pH from positive to negative and to render the adjacent peptide bonds resistant to hydrolysis by trypsin; it can also lead to disaggregation of a multimeric protein. The reaction occurs at pH 8.5 (usually at 2°C) and the maleyl groups can be removed at pH 3.5 (usually at 60°C). Maleylation of thiol groups may also occur.

Malonate Semialdehyde

It is the recommended trivial name for malonate semialdehyde; formylethanoate; OHC–CH2–COO– . It is an intermediate in the oxidation of propionyl-CoA to acetyl-CoA in the terminal stages of the beta-oxidation of fatty acids with an odd number of carbon atoms in plants.

Maltoporin

A bacterial outer membrane transmembrane protein, with possibly 16 helices, having two roles in *Escherichia coli*: it transports maltodextrins (including maltose) into the cell, and acts as a receptor for lambda and some other phages.

Mammal

Any animal of the Mammalia, a large class of warmblooded tetrapod vertebrates characterized by the possession of sweat glands in the skin and, generally, an insulating layer of hair. Female mammals characteristically suckle their young at modified sweat glands – mammary glands – that secrete milk. Like birds, mammals possess a four-chambered heart, but their thoracic diaphragm is a uniquely mammalian feature.

Mannan

The main hemicellulose of soft (coniferous) wood, made up of D-mannose, D-glucose, and D-galactose in the ratio of 3:1:1. Glucomannans occur in ferns and in certain higher plants and algae where they may function as the main structural component of the cell wall. Highly branched mannans can be extracted from the cell walls of some yeasts.

N

Nalidixic Acid

Nalidixic acid 1-ethyl-7-methyl-1,8-naphthyridin-4-one-3-carboxylic acid; a synthetic 4-quinolone antibiotic that is active against many Gram-negative bacteria. It specifically inhibits bacterial DNA gyrase thereby interfering with DNA synthesis. It has no apparent effect on eukaryotic DNA synthesis.

Natriuretic-Peptide Receptor

Natriuretic-peptide receptor any type 1 membrane glycoprotein that binds natriuretic peptides and mediates their intracellular effects. Three receptors have been identified: ANPA (previously known as GC-A, ANP-R1 , or ANPB); ANPB (previously known as GC-B); and ANPC. Human ANPA and ANPB each have one transmembrane domain and intrinsic guanylate cyclase activity in an intracellular domain, which is highly conserved (88% identity). At the ANPA receptor, atrial natriuretic peptide is about ten times more potent than brain natriuretic peptide, whereas type C natriuretic peptide is 50–500 times more potent than other agonists at the ANPB receptor. ANPC is a disulfide-linked homodimer with one transmembrane domain and no guanylate cyclase domain; it acts through other second messenger systems.

Natural Language Processing

Natural language processing abbr.: NLP; the computational analysis and interpretation of human language. NLP is used in software that provides automatic translations of text from one language to another, in robotic systems that use human-language-type commands, and in text-mining tools (e.g. to provide summaries or abstracts of large volumes of text).

Natural Selection

The principle that the best competitors in any given population of organisms have the best chance of breeding success and thus of transmitting their characteristics to subsequent generations. The members of any population show individual differences – anatomical, physiological, or metabolic – that affect their functional efficiency in a given environment. The less efficient members tend to die out or produce fewer offspring than the more efficient members, which are better adapted to compete for food or other resources, and so produce relatively more offspring. The principle is fundamental to modern concepts of evolution, and was first articulated, independently, by the British naturalists Alfred Russel Wallace (in 1858) and Charles Darwin (in 1859).

Natural-Selection Theory of Immunity

Natural-selection theory of immunity or Jerne's theory of immunity a theory of antibody production that postulates the spontaneous presence, in the blood of an animal, of small numbers of antibody molecules against all antigens to which the animal can respond, and delegates to the antigen the sole role of carrying such specific globulin molecules from the bloodstream into cells in which these molecules can induce proliferation of a particular antibody. Nb symbol for niobium.

ncRNA

ncRNA abbr. for noncoding RNA; strictly, any RNA that is not mRNA, but the term also commonly excludes tRNA and rRNA. The term is used predominantly in relation to eukaryotes, with sRNA (small RNA) being used in relation to prokaryotes. It encompasses many varieties of RNA that have specific, but noncoding, functions. Examples are: miRNA (micro RNA; 21–25 nt), which modulates development in many eukaryotes; sRNA (≈100–200 nt), which are translation regulators in bacteria; 7S RNA (300 nt) of the signal recognition particle; U-snRNA (100–215 nt), which are involved in spliceosomes in eukaryotes; and RNAs (>10 000 nt) that are involved in gene silencing in higher eukaryotes.

Nebulin

A giant protein (800 kDa) that is specific for the skeletal muscle of vertebrates; it acts possibly as a 'molecular ruler' for length regulation of the thin filament. The protein appears to have small repeats of approximately 35 amino acids.

Necrosis

The death of a portion of a tissue as a result of disease or injury. The cells swell, their plasma membranes become disrupted, and the cell contents are released into the extracellular space, where they often trigger an inflammatory response. The process is unregulated and should be distinguished from apoptosis.

Nedd

Nedd abbr. for NPC-expressed, developmentally down-regulated genes (NPC is an abbreviation of neural precursor cells). These genes are down-regulated during mouse brain development. The products are proteins presumed to be involved in embryonic development and differentiation of the nervous system. Nedd-2 protein is a cysteine endopeptidase that may be involved in controlling cell death. Nedd-4 is ubiquitously expressed in mammals. It contains multiple WW domains, which bind with great affinity to the b and c subunits of epithelial Na channel (ENaC), and a ubiquitin-ligase domain. Nedd-8 (called Rub 1 in yeast) is a ubiquitin-like molecule in mammals that becomes conjugated with cullin 1 and promotes its degradation in proteasomes.

Neddylation

The process whereby Nedd-8 (or Rub 1 of yeast), a ubiquitin-like protein conjugate to some proteins to target them for degradation by proteasomes or for deneddylation by COP signalosomes. It is important in controlling the auxin response pathway in plants.

Needleman–Wunsch Alignment Algorithm

A dynamic programming algorithm used in sequence alignment that employs a 2D search matrix, such that its speed and storage requirements are of the order N × M when sequences of length N and M are aligned. The algorithm has large memory requirements owing to its traceback step, where the matches of the optimal alignment are reconstructed, making it too slow for routine database searches

Negative-Contrast Technique

A technique used in preparing specimens for electron microscopic examination in which the specimen is mixed with an electron-dense material that penetrates the interstices of the specimen but not the material of the specimen itself. The specimen thus appears transparent against an opaque background. It is commonly called negative staining

Negative-Strand Virus

Negative-strand virus or negative-stranded virus any RNA virus in which the genome consists of single-stranded RNA (i.e. minus strand) of base sequence complementary to that of the virus-specified mRNA (which is the positive, or plus, strand). Such viruses, which comprise class V in the Baltimore classification, must make a positive RNA strand before making viral proteins

Neighbour-Joining

A method for inferring phylogenetic trees based on the principle of minimum evolution. The method resolves the phylogeny in a stepwise fashion by selectively joining pairs of taxa that minimize the sum of branch lengths. Neighbour joining is computationally efficient and is an appropriate reconstruction approach where evolution has not proceeded in a strictly clocklike manner.

Nemaline Myopathy

abbr.: NEM; a slowly progressing myopathy characterized by rodlike structures, composed largely of a-actinin and actin, in skeletal muscle fibres. Nem1 (gene locus at 1q22-q23) is an autosomal dominant disease and is associated with a missense mutation in a-tropomyosin. NEM2 (gene locus at 2q22) is an autosomal recessive disease whose gene locus is that for nebulin.

Nematic

Nematic (of a substance) being in or having a mesophilic state in which the molecules are arranged as in a series of parallel threads but not in layers. The molecules can rotate about their axes and can move in the plane orthogonal to the line of the thread. Compare smectic.

Neomycin

An aminocyclitol antibiotic complex produced by *Streptomyces fradiae*. It consists of: neomycin A (or neamine; a degradation product of neomycin B and neomycin C), neomycin B, and neomycin C. Neomycin causes misreading during protein synthesis in bacteria but its target site on the ribosome is apparently different from that of streptomycin. Neomycin resistance is conferred by the enzyme aminoglycoside 3′-phosphotransferase.

Neonatal Jaundice

A physiological predominantly unconjugated hyperbilirubinemia with clinical jaundice, affecting about half of all human neonates during the first five days of life. It results from increased bilirubin production and delayed maturation of liver UDP-glucuronosyltransferase activity. Maternal–fetal Rh blood group incompatibility and hereditary hyperbilirubinemia syndromes exaggerate the condition, which, if untreated, can lead to kernicterus and brain damage.

Nernst Distribution Law

When, at a constant temperature, a solute distributes itself between two immiscible phases, then the ratio of its concentrations in the two phases is constant, and is described by the relation $C1 / C2 = K$, where C1 and C2 are, respectively, the amount of-substance concentrations of the solute in phases 1 and 2 at equilibrium, and K is the distribution constant. The law only applies to dilute solutions. [After Walther Hermann Nernst (1864–1941), German physical chemist.]

Nephelometry

A technique in which the light dispersed by a suspension is measured orthogonally to the direction of the incident light; the amount of scattered light is dependent on the number and size of the particles in the light path.

Nephrocalcin

Nephrocalcin or osteocalcin-related protein precursor a calcium-binding protein found in bone. It is expressed only in the kidney; in humans and other mammals, it is also found as a urinary acidic glycoprotein that strongly inhibits the formation of calcium oxalate crystals in renal tubules. It contains c-carboxyglutamic acid residues; a modified protein lacking these is found in patients with calcium oxalate renal .calculi

Nephrogenic Diabetes Insipidus

A rare X-linked recessive genetic disorder involving renal insensitivity to vasopressin resulting in polyuria and polydipsia; it can also be acquired as a result of toxic damage to the tubules. The genetic form is associated with mutations in the V2 receptor (see vasopressin receptor), two examples being the substitution of Asp for Ala at position 132 and a frameshift mutation at position 246 leading to a premature stop codon and a truncated receptor.

Nephron

Any of the structural and functional urine-secreting units that occur in the kidney. Typically each nephron is made up of a glomerulus, a proximal convoluted tubule, a loop of Henle, and a distal convoluted tubule.

Nerve

Any of the cordlike bundles, consisting of nerve fibers and glia encased in a connective-tissue sheath, that connect the central nervous system to other parts of the body. They may be motor, sensory, or of mixed function.

Nerve Fibre

The axonal process of a neuron together with its covering sheath. Bundles of nerve fibers running together make up a nerve.

Nerve Gas

Any of various poisonous gases or volatile liquids that act by inhibiting the passage of impulses through the nervous system and neuromuscular junctions. They are commonly irreversible inhibitors of acetylcholinesterase.

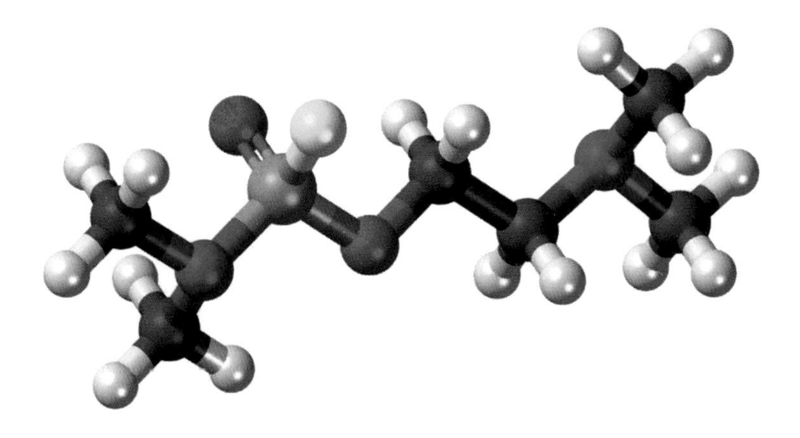

Figure 29. An Illustration of Nerve Gas Agent.

Source: Image by Wikimedia Commons

Nerve Growth Factor 1

Any of a number of polypeptides that exert a trophic effect on neurons. They play a part in the development and maintenance of sensory neurons in dorsal root ganglia and sympathetic neurons in peripheral sympathetic ganglia. 2 abbr.: NGF; the first known member of a family of polypeptides that act as growth factors for neurons. In addition to the general properties of such nerve growth factors, it also stimulates growth and differentiation of B lymphocytes, and stimulates histamine release from mast cells. The biologically active form of NGF contains two identical polypeptide chains of 120 amino-acid residues and known sequences and exerts its action through specific receptors in the neuronal plasma membrane. It is produced by neurons, astrocytes, and Schwann cells, but also by fibroblasts, epithelial cells, activated macrophages, and smooth muscle cells. It exists as an inactive complex of two a subunits, two b subunits, and two c subunits. The active form consists of the two b subunits, and is known as bNGF. Related proteins subsequently discovered include brain-derived neurotrophic factor and neurotrophins.

Nerve Growth Factor

Receptor any of a family of plasma membrane integral proteins that bind nerve growth factors. Low-affinity and high-affinity binding is observed; high affinity binding, and action on effector systems, require the presence of both the trk protooncogene product (p140prototrk) and a 75 kDa low-affinity receptor glycoprotein (p75NGFR). The latter can bind nerve growth factor (def. 2), brain-derived neurotrophic factor, and neurotrophins 3 and 4, but it is apparent that considerable specificity exists in the transduction of the signal depending on which member of the nerve growth factor family and which cell type is involved.

Nervous System

The extensive network of cells specialized to carry information, in the form of nerve impulses, to and from all parts of the body. Nervous systems occur in all orders of multicellular animals other than sponges.

Nesidioblastosis

A pathological condition in which individual blastic pancreatic duct cells differentiate abnormally into islet cells scattered among the exocrine tissue. The condition is sometimes associated with severe infantile hyperinsulinism and hypoglycemia.

Nessler's Reagent

An alkaline aqueous solution of potassium tetraiodomercurate(ii). It gives a yellow or brown color or precipitate when added to solutions containing ammonia or ammonium ions and is useful in the detection and quantitative estimation of microgram quantities of ammonia in aqueous solution.

Nested PCR Primers

A set of oligonucleotide primers used for the amplification of DNA by the polymerase chain reaction (PCR) in which the outermost 5′ and 3′ pair are used in the first phase of amplification and a second pair is designed to prime within that PCR product to produce a shorter amplified sequence. Greater specificity of amplification is expected from this use of two pairs of primers.

Nestin

An intermediate-filament protein whose name derives from neuroepithelial stem cells, in which it is specifically expressed. It shares significant similarities with other intermediate-filament proteins, including a set of heptad repeats.

Neu

An oncogene first described in association with a neuroblastoma after treatment of rats with ethylnitrosourea. The neu protooncogene (also known as c-erbB2, NGL, and HER2) encodes a 185 kDa transmembrane glycoprotein with tyrosine kinase activity, related to but distinct from the epidermal growth factor receptor (EGFR). The neu gene product, Neu, possesses a cysteine-rich extracellular domain, a transmembrane domain, and an intracellular tyrosine kinase domain. EGF does not bind to Neu, but a factor known as Neu-activating factor (NAF) binds specifically to Neu and will bind to EGFR. Neu was originally identified in brain, and may play a critical role in neurogenesis.

Neuberg's Second Form of Fermentation

The anaerobic fermentation of carbohydrates by yeast in the presence of sodium hydrogen sulfite. the acetaldehyde formed normally, which would act as a hydrogen acceptor for the reoxidation of NADH, is trapped as its bisulfite addition product, and its place is taken by dihydroxyacetone phosphate (i.e.glycerone phosphate), which is reduced to snglycerol 3-phosphate; this is then dephosphorylated to glycerol. Thus one mole each of glycerol, acetaldehyde–bisulfite complex, and carbon dioxide is formed per mole of glucose fermented.

Neuberg's Third Form of Fermentation

The anaerobic fermentation of carbohydrates by yeast in alkaline conditions. The acetaldehyde formed normally, which would act as a hydrogen acceptor for the reoxidation of NADH, undergoes a Cannizzaro reaction to form acetic acid and ethanol, and its place is taken by dihydroxyacetone phosphate (e.g.glycerone phosphate), which is reduced to sn-glycerol 3-phosphate; this is then dephosphorylated to glycerol. Thus one mole each of glycerol and carbon dioxide, and one-half mole each of ethanol and acetic acid, are formed per mole of glucose fermented.

Neuregulin

Any member of a family of closely related proteins implicated as regulators of neural and muscle development, and differentiation and oncogenic transformation of mammalian epithelia. They include the Neu differentiation factor (see neu) and the heregulins. They exert their effects through binding to and activating, receptors of the ErbB2/Neu receptor family. Neuregulins stimulate the proliferation of Schwann cells, increase the rate of synthesis of acetylcholine receptors in cultured muscle cells, and are concentrated at nerve–muscle synapses.

Neurofibromatosis

Any of a number of disorders associated with the presence of multiple neurofibromas, benign tumors consisting of a mixture of Schwann cells and fibroblasts. Type 1 is a common autosomal dominant disorder that results from mutations in neurofibromin. Type 2 is also autosomal dominant but rare and results from mutations in merlin (also called schwannomin).

Neurotensin

A tridecapeptide, Glp-Leu-Tyr-Glu-Asn-Lys-Pro- ArgArg-Pro-Tyr-Ile-Leu-OH, found in mammalian brain and gut, especially in N cells, packaged within secretory vesicles. It is released from the hypothalamus into the circulation. Although initially shown to induce hypotension, it is now known to have a wide variety of pharmacological actions, including muscle contraction. The precursor gives rise to neurotensin and two other derivatives, neuromedin and neuromedin N.

Neuromuscular Junction

The place of contact between the motor end-plate of the fiber of a motor neuron and the membrane of a muscle fiber supplied by the neuron. Impulses are transmitted across the gap by the diffusion of a neurotransmitter.

Oakley–Fulthorpe Technique

A simple double immunodiffusion method that can be used for the quantitative determination of antigen concentrations; it is essentially a development of the Oudin technique. The antiserum in 1 agar is placed at the bottom of a cylindrical tube above which is placed a layer of 1 agar gel in 1 saline, and 0.5 o-cresol (as a preservative). Subsequently, a bacterial filtrate (containing antigen) is layered onto the agar–saline gel and the tube is then incubated at 37 °C. Disks or lines of flocculation of antigen–antibody complexes are produced in the agar–saline gel, from whose positions the concentration of antigen(s) may be determined.

Obese

Having an excessive amount of body fat. Obesity is the second largest cause of preventable death in the developed world and a major risk factor for vascular disease, diabetes mellitus type 2, and some cancers. —obesity

Obestatin

A peptide encoded by the ghrelin gene that opposes ghrelin's effects on food intake. The 23-residue peptide, has the structure FNAPFDVGIKLSGAQYQQHGRAL-amide in the rat. Treatment of rats with the peptide depresses food intake, inhibits jejunal contraction and results in decreased body weight gain. Ghrelin and obestatin are both derived from the ghrelin gene propeptide by posttranslational cleavage and modification. Obestatin binds to the orphan G-protein-coupled receptor GPR39.

OB Fold

abbr. for oligonucleotide/oligosaccharide binding fold; a protein sequence (≈110 residues) present in most prokaryotic and eukaryotic ssDNA-binding proteins. It does not have a definite sequence motif but forms a highly curved five-stranded beta-sheet that closes in on itself to form a beta-barrel. The BRCA2 protein contains three OB folds.

Object-Oriented Database

abbr.: OODB; a database in which data are stored as abstract objects, with abstract relationships between them. The data representations are flexible (e.g. including character strings, digitized images, tables) and objects may be grouped together.

Oblate 1

Having an equatorial diameter longer than the polar diameter. 2 describing an ellipsoid of rotation (including a spheroid) generated by rotation of an ellipse about its minor axis. 3 describing any protein of oblate (def. 1) dimensions, as observed in hydrodynamic measurements. Compare prolate.

Obligate

Obligate by necessity, or without option; used especially of the nature of the environment or of the mode of life of an organism; e.g. obligate aerobe (see aerobe), obligate anaerobe (see anaerobe). Compare facultative.

OB/OB

The genotype symbol for any strain of mouse with an inherited form of severe obesity, due to a homozygous autosomal recessive mutation on chromosome 6. Such animals display marked hyperglycemia and insulin resistance. A protein named leptin, or OB protein, has been implicated in the control of fat mass and is a product of the ob mutant gene, being deficient in ob/ob mice.

Occult Blood

Blood that is present in a sample in such small quantities, e.g. in feces, that it can only be detected by chemical testing or by microscopy

Occupancy 1

The state or condition of being filled or occupied. 2 the number of similar sites occupied, usually expressed as a fraction or percentage of the total number available. It can refer, e.g., to the number of receptors on a cell surface occupied by a particular agonist or to the number of atomic orbitals of a particular subshell filled by electrons

Occupation Theory of Agonist Action

A theory stating that the magnitude of the response to an agonist is directly proportional to the fraction of specific receptors occupied by molecules of the agonist; it assumes that the binding of agonist to receptor is reversible. According to this theory, the response will increase to a maximal value when all the receptors are occupied. Compare rate theory of agonist action.

Ochratoxin

Any one of several mycotoxins produced by the fungus *Aspergillus ochraceus*, other *Aspergillus* spp., and *Penicillium* spp.; the major component, ochratoxin A, is a complex, chlorine-containing derivative of L-phenylalanine that inhibits phosphoenolpyruvate carboxykinase (EC 4.1.1.49) and may cause fatty liver. These toxins may occur on contaminated foodstuffs such as corn (maize), peanuts, storage grains, etc.

Ochre Suppressor

Any of a number of mutations in *E. coli* resulting in changed anticodons of tRNAs that suppress an ochre codon (UAA) in mRNA. This allows the insertion of one of several alternative amino acids into a polypeptide at that site. Examples are supC (Tyr) and supG (Lys). Some ochre suppressors (supC and supG) also suppress amber codons.

Octadecadienoic Acid

Any straight-chain fatty acid having eighteen carbon atoms and two double bonds per molecule. Linoleic acid (see linoleate) is the all-Z-(9,12)-isomer and is a constituent of most vegetable oils and animal fats. The all-E-isomer of this acid is linolelaidic acid, a constituent of the seeds of *Chilopsis linearis*, which also contains the all-E-(10,12)-isomer. A number of other isomers have been synthesized chemically.

Octadecatetraenoic Acid

Any straight-chain fatty acid having eighteen carbon atoms and four double bonds per molecule. Several isomers occur naturally, including the 3E,9Z,12Z,15Z-isomer, found in seed oil, and the all-Z-(6,9,12,15)-isomer, found in fish oil. The 9Z,11E,13E,15Z-isomer and the all-E-(9,11,13,15)-isomer are a- and b-parinaric acid respectively, from *Parinarium laurinum*; both isomers are used as fluorescent probes.

Octahedron (pl. octahedra)

Any solid geometrical figure having eight plane triangular faces, 12 edges, and six tetrameric vertices. In a regular octahedron, the faces are congruent equilateral triangles and may be considered to consist of two equal pyramids opposing each other on the same square base. —octahedral adj.

Octamer 1

An eight-base sequence element that is common in eukaryotic promoters. The consensus sequence is ATTTGCAT. It binds a number of transcription factors (see OCT). 2 any polymer of a protein having eight subunits, e.g. b-lactoglobulin at pH 4.7 below 4°C. 3 an assembly of eight histones containing two each of H2A, H2B, H3, and H4, that forms the histone core of the nucleosome.

Octopamine

b-hydroxytyramine; 1-(p-hydroxyphenyl)-2-aminoethanol; the D(–)-enantiomer is a biogenic amine, about one-tenth as active as norepinephrine, formed by the b-hydroxylation of tyramine by dopamine b-hydroxylase. It is found in the salivary glands of *Octopus* spp. and of *Eledone moschata*.

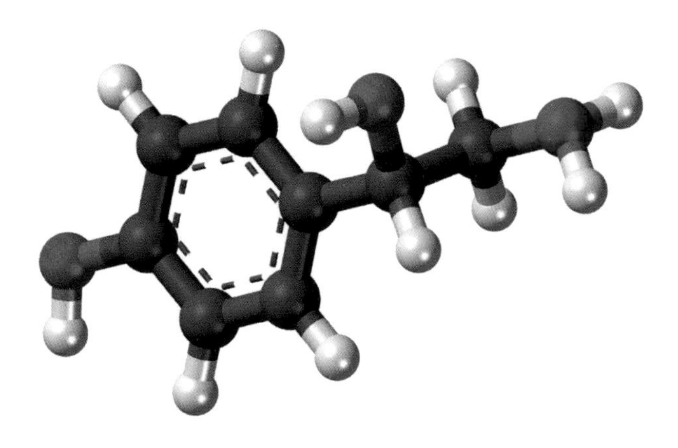

Figure 30. An Octopamine Molecule Structure.

Source: Image by Wikimedia Commons

Octopine 1

N-(1-carboxy-4-guanidinobutyl)-L-alanine; N2 -(1-carboxyethyl)-L-arginine; D-octopine (i.e., N2 -(D-1-carboxyethyl)-Larginine) is an opine found in the tumors of crown-gall disease in plants and in the muscles of certain invertebrates. The genes responsible for the synthesis of octopine are part of the T-DNA from a Ti-plasmid. 2 any opine that, like octopine itself, is an N-substituted derivative of alanine. Octopine was first isolated from the muscles of Octopus, but is found in other cephalopod species and lamellibranchs. The guanidine group can undergo phosphorylation to phosphooctopine, which acts as a phosphagen. Compare octopinic acid.

Octopinic Acid

N-(1-carboxy-4-aminobutyl)-L-alanine; N2 -(1-carboxyethyl)-L-ornithine; D-octopinic acid (i.e., N2 -(D-1-carboxyethyl)- L-ornithine) is an opine found in the tumors of crowngall disease of plants; it is a member of the octopine (def. 2) family.

Octose

Any aldose having a chain of eight carbon atoms in the molecule

Octreotide

A synthetic octapeptide analog of somatostatin, with which it shares four residues essential for activity. It has a much longer half-life in vivo (1–2 h) than somatostatin (1–2 min) and has been used in the therapy of breast, ovarian, prostatic, gut, endocrine, and pituitary tumors

Octyl Glucoside

n-octyl b-D-glucopyranoside; a mild nonionic detergent, aggregation number 84, CMC 20–25 mm. The a-glucoside isomer has also been used, e.g. for crystallization of membrane proteins. The lack of absorbance at 228 nm is an advantage in the use of these compounds.

Odorant-Binding Protein

Any of a family of some 10 proteins that bind and transport odorant molecules across the mucus layer that covers odour receptors in the nasal mucosa. They are homodimers that contain an odorant-binding domain and a dimerization region.

Odorant Receptor

Any G-protein-coupled membrane receptor that binds and elicits the biological response of an odorant molecule. Several hundred genes for such receptors are present in the human genome. The receptors function through Gaolf, which activates adenylate cyclase, or GaP, which activates phospholipase C. They are present in cells of the olfactory system and also outside it (there are over 50 different receptors in spermatogenic cells).

OGCP

abbr. for oxoglutarate/malate carrier protein; a mitochondrial inner membrane integral protein that plays an important role in several processes, including the malate– aspartate shuttle, gluconeogenesis from lactate, and nitrogen metabolism.

Ogston Concept or Three-Point Attachment Hypothesis

A concept formulated to explain the inherent differing reactivities, often expressed in an enzymic reaction, of identical chemical groups in a prochiral molecule. It states that there must be at least three points of attachment of a substrate molecule to the active site of an enzyme. It is similar to the

Easson–Stedman model. [After Alexander George Ogston (1911–96), British biophysicist who formulated it in 1948.]

Ohm

the SI derived unit of electric resistance, defined as the resistance between two points on a conductor through which a current of one ampere flows as the result of a potential difference of one volt applied between the points, the conductor not being the source of any electromotive force; i.e. 1 X = 1 VA–1. [After Georg Simon Ohm (1787–1854), German physicist.]

Ohm's Law

A law stating that under constant conditions the current, I, flowing through a given conductor is proportional to the potential difference, U, applied across it. The law is often expressed in the form U = IR, where the proportionality constant, R, is the resistance of the conductor.

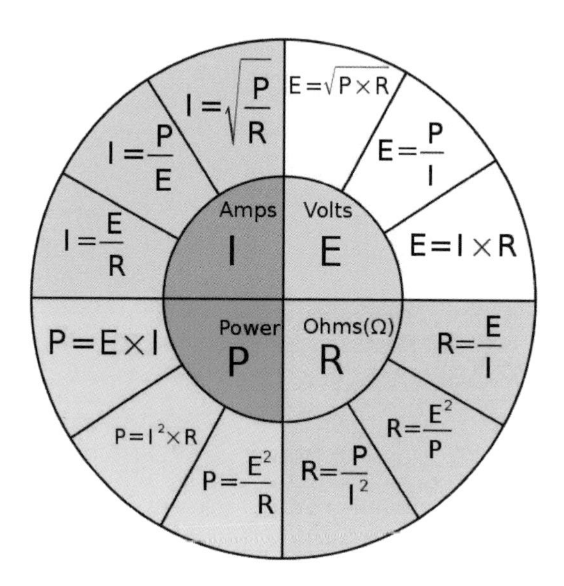

Figure 32. Ohm's Law Pie Chart.

Source: Image by Wikimedia Commons

+OID

Suffix forming adjectives and (associated) nouns 1 denoting likeness to, or having the form of (something specified); e.g. colloid (i.e. gluelike substance). 2 generated by rotation of (the geometrical figure specified); e.g. ellipsoid (i.e. solid figure derived from an ellipse). 3 belonging to the class or group represented by (the type member specified); e.g. steroid (i.e. compound belonging to the same chemical family as sterols). —+oidal adj

Oil

Any neutral, flammable substance that is liquid at room temperature and is characteristically soluble in relatively nonpolar solvents but only sparingly soluble in aqueous solvents. There are three main groups: (1) animal and vegetable oils, which usually consist predominantly of triacylglycerols but may contain varying amounts of fatty-acid esters of other alcohols; (2) mineral oils, derived from petroleum, coal, shale, etc., which consist predominantly of hydrocarbons; and (3) essential oils.

Oil-Immersion Objective

An objective lens in a light microscope designed to be used with a layer of a special oil between the object (or its coverslip) and the lens. The oil is of the same refractive index as the glass of the lens. This arrangement maximizes the numerical aperture and hence the resolving power of the lens.

Okadaic Acid

A polyether fatty acid, Mr = 804.9, is a potent inhibitor of protein phosphatases, especially the PP-1 and PP-2A types. It is a tumor promoter, and is implicated as the causative agent of diarrhetic shellfish poisoning. Its name derives from *Hilichondria okadaii*, a marine sponge that feeds on *Prorocentrum lima*, a dinoflagellate from which it is isolated.

Okazaki Fragment or Okazaki Piece or Okazaki Segment

Any of the relatively short polydeoxyribonucleotides that are formed, concomitantly with the continuous replication of one of the two strands of a duplex DNA molecule, as intermediates during the discontinuous replication of the other strand. Such fragments are synthesized from mononucleotides in a direction opposite to that of the movement of the replication fork; then, as replication proceeds, they are covalently joined through the action of a polynucleotide ligase to form a long daughter polynucleotide chain, the lagging strand. These fragments appear to be 1000–2000 residues long in *Escherichia coli* and 100–200 residues long in mammalian cells; a proportion of them may have RNA primer attached.

Oleic Family (of Polyunsaturated Fatty Acids)

A series of polyenoic acids in which the hydrocarbon chain terminates with the alkenyl grouping CH3–[CH2]7–CH=CH– (as in oleic acid); there are three other series, i.e. the linoleic family, linolenic family, and palmitoleic family. The oleic family can be synthesized from acetyl-CoA: palmitoyl-CoA is first synthesized by the fatty-acid synthase (see fatty acid synthase complex); next, by chain elongation and action of desaturases, stearoyl-CoA, oleoyl-CoA, and longer chain fatty acyl-CoA compounds are synthesized; then from their respective CoA derivatives, the cognate fatty acids are formed.

Oligodendrocyte

A type of glial cell that forms and supports the myelin sheath around axons in the central nervous system of vertebrates. In contrast to the Schwann cell, which performs this role in peripheral neurons for a single axon, one oligodendrocyte is involved in the myelination of several axons.

Oligo-1,6-Glucosidase

(EC 3.2.1.10), other names: sucrase-isomaltase; limit dextrinase; isomaltase; an enzyme that catalyzes the hydrolysis of 1,6-a-D-glucoside linkages in isomaltose and dextrins produced from starch and glycogen by a-amylase. A similar enzyme (EC 3.2.1.48) is a type 2 transmembrane protein of the human intestinal brush border. This contains two catalytic domains that share 40 homology. Congenital sucrase-isomaltase deficiency is rare and is genetically heterogeneous.

Oligomer 1

(In chemistry and biochemistry) Any substance or type of substance that is composed of molecules containing a small number – typically two to about ten – of constitutional units in repetitive covalent linkage; the units may be of one or of more than one species.

Oleosin

Any of a family of integral membrane proteins (15–25 kDa) of the endoplasmic reticulum in cells of oilseeds that define the accumulation of triacylglycerols and formation of oil bodies (or spherosomes). Oleosins each consist of a conserved central hydrophobic region (70–80 residues) flanked by hydrophobic termini.

Olfactory Receptor

Any integral membrane protein thought to be involved in affecting the sense of smell. They are G-protein-associated receptors, and the human genome contains several hundred genes for such receptors.

Oligodendrocyte

A type of glial cell that forms and supports the myelin sheath around axons in the central nervous system of vertebrates. In contrast to the Schwann cell, which performs this role in peripheral neurons for a single axon, one oligodendrocyte is involved in the myelination of several axons.

Oligonucleotide

abbr. (colloquial): oligo; any molecule that contains a small number of nucleotide units connected by phosphodiester linkages between (usually) the 3′ position on the glycose moiety of one nucleotide and the 5′ position of the glycose moiety of the adjacent one. The number of nucleotide units in these small singlestranded nucleic acids (usually DNA) is variable but often in the range of 6 to 24 (hexamer to 24mer), although for some purposes it might be as long as 50mer. They are widely used as primers for reactions catalysed by DNA polymerases.

Oligotrophic

Describing a habitat, especially a lake or other mass of water, that is poor in nutrients capable of supporting the growth of aerobic plants and microorganisms, and (in the case of a lake, etc.) that hence contains abundant dissolved dioxygen at all depths.

Omeprazole

The nonproprietary pharmaceutical name for 5-methoxy2- [(4-methoxy-3,5-dimethyl-2-pyridinyl)methyl]sulfinyl-1H-benzimidazole, a gastric proton-pump inhibitor. It blocks both basal and stimulated secretion of acid in the stomach by reversibly inhibiting H+/K+-transporting ATPase (EC 3.6.1.36) in the plasma membranes of parietal cells within the oxyntic glands of the gastric mucosa. It is useful clinically in the treatment of erosive reflux esophagitis, benign peptic ulcer, and Zollinger–Ellison syndrome. Proprietary name: Losec.

Ommochrome or Ommatochrome

Any yellow, brown, red, or violet natural polycyclic pigment especially common in the Arthropoda, particularly in the ommatidia of the compound eye. Ommochromes contain a phenoxazine nucleus and are formed biologically from tryptophan via kynurenine. They are divided into two groups: ommatins, which are alkali-labile, of low relative molecular mass, and rather weakly colored; and ommins, which are alkali-stable, of higher relative molecular mass, and more strongly colored than the ommatins.

+Onium

Noun suffix denoting a cationic molecular entity in which a heteroatom is bonded to one more univalent ligand than is normal for a neutral molecule containing that heteroatom and therefore bears a formal positive charge; e.g. ammonium, H4N+; dimethylammonium, $(CH_3)2H_2N+$; diphenyliodonium, $(C_6H_5)_2I^+$; trialkylsulfonium, R_3S + (where R = alkyl).

P

PACAP

abbr. for pituitary adenylate cyclase-activating polypeptide; a preganglionic neuropeptide that stimulates expression of genes for chromogranins A and B (all of which have a cAMP response element in their promoter) in neuroendocrine cells. When released by melanopsin-containing retinal axons it transmits photic information.

Packaging Extract

A lysate of *E. coli* in which all but one of the capsid and tail proteins of bacteriophage lambda have been expressed. A mixture of two such lysates lacking different proteins is capable of packaging recombinant lambda genomes with high efficiency *in vitro*.

Packaging Cell Line

Any mammalian cell line modified for the production of recombinant retroviruses. They express essential viral genes that are lacking in the recombinant retroviral vector.

Packing

(In chromatography) the material (whether active solid, liquid held on solid support, or swollen gel) that is introduced into a column and that consists of or contains what is to become the stationary phase during a chromatographic separation.

Packing Density

The ratio of the minimum (or actual) volume of an object, such as a molecule in a crystal or in solution, to the total volume it occupies. For a macromolecule or region of a macromolecule, it is the ratio of the volume enclosed by the van der Waals envelope of all atoms in the molecule or region to the total volume of the molecule or region. Theoretical values are 0.74 for close-packed spheres, 0.91 for infinite cylinders, and 1.0 for a continuous solid.

Paired

The gene for a sequence-specific DNA-binding homeobox segmentation protein in Drosophila melanogaster. It is one of a number of pair-rule genes, each of which specifies a simple alternation with a repeat distance of two segments. Their protein products are characterized by a paired-box DNA-binding domain of ≈128 amino acids, which consists of two subdomains each resembling a homeobox domain.

PaJaMa Experiment

The nickname for a bacterial mating experiment carried out at l'Institut Pasteur, Paris that showed that induction and repression of the enzyme b-galactosidase in *Escherichia coli* are regulated by two closely linked genes, one of which produces a cytoplasmic repressing substance that blocks the expression of the other.

PAK

abbr. for p21-activated protein kinase; any of a family of protein serine/threonine kinases activated by p21 and Cdc42/Rac (a Rho GTPase involved in regulation of the actin cytoskeleton), that are differentially expressed in mammalian cells. PAKs are upstream modulators of protein kinase cascades involved in regulation of apoptosis and in morphogenesis. They have an N-terminal p21- Cdc42/Rac-binding domain and a C-terminal catalytic domain. PAK1 localizes with actin filaments of the cytoskeleton. PAK2 is cleaved into its two domains by certain caspases. A mutation in PAK3 results in an X-chromosome-linked form of mental retardation

Palindrome 1

Any linear arrangement of symbols, such as letters or digits, that has the same sequence from either end; e.g. noon, radar; 75311357; 'Madam, in

Eden I'm Adam'. 2 (in molecular biology) palindromic sequence a DNA sequence with a twofold rotational axis of symmetry (dyad symmetry). Confusingly the term is used in at least two senses: (1) to describe a region of local twofold rotational symmetry in duplex DNA.

Pallidin

Any of a group of endogenous carbohydrate-binding proteins (lectins) produced by cells of the slime mold *Polysphondylium pallidum* during differentiation. Similar to discoidin, they may be involved in cell adhesion.

Palmitate 1

Numerical symbol: 16:0; the trivial name for hexadecanoate, CH3–[CH2]14–COO– , the anion derived from palmitic acid (i.e.hexadecanoic acid), a saturated straight-chain higher fatty acid. 2 any mixture of free palmitic acid and its anion. 3 any salt or ester of hexadecanoic acid.

Palmitoleate 1

Numerical symbol: 16:1(9); the trivial name for (Z)- hexadec-9-enoate, CH3–[CH2]5–CH=CH–[CH2]7 -COO– , the anion derived from palmitoleic acid, (Z)-hexadec-9-enoic acid, a monounsaturated straight-chain higher fatty acid. 2 any mixture of free palmitoleic acid and its anion. 3 any salt or ester of 9-hexadecenoic acid

Palmitoleic Family (of Polyunsaturated Fatty Acids)

A series of polyenoic acids in which the hydrocarbon chain terminates with the alkenyl grouping CH3–[CH2]5–CH=CH– (as in palmitoleic acid (see palmitoleate)); there are three other series of such acids, i.e. the linoleic family, linolenic family, and oleic family. Members of the palmitoleic family can be synthesized from palmitic acid via palmitoleic acid by chain elongation and/or desaturation, but in mammals not from linoleic, (9,12,15)-linolenic, or oleic acids; the series includes cis-vaccenic acid.

Palmitoleoyl

The trivial name for (Z)-hexadec-9- enoyl, CH3–[CH2]5–CH=CH–[CH2]7–CO– (cis isomer), the acyl group derived from palmitoleic acid (i.e. (Z)-hexadec-9-enoic acid). It is a relatively minor component of plant and animal lipids. The chain is synthesized metabolically by the action of a D9 -desaturase on palmitoyl-CoA.

Palmitoyl

symbol: Pam; the trivial name for hexadecanoyl, CH3–[CH2]14–CO–, the acyl group derived from palmitic acid (i.e.hexadecanoic acid). This is one of the major fatty-acid components of plant and animal lipids; together with stearic acid, it represents a high proportion (often around 30%) of the fatty-acid content of dietary and other lipids. It acts as a precursor in mammals for a family of unsaturated fatty acids in which all double bonds are nine or more carbons from the methyl terminus. The chain is synthesized metabolically from acetyl-CoA as palmitate (in mammals) or palmitoyl-CoA (in yeast), the product of fatty-acid synthase.

Palmitoyl-Protein Thioesterase

EC 3.1.2.22. a lysosomal enzyme that cleaves the thioester linkage formed by a palmitoyl moiety with a cysteine side chain near a transmembrane segment of an integral membrane protein. Mutations in this gene on 1p32 cause infantile neuronal lipofuscinosis and progressive myoclonic epilepsy.

Palytoxin

Any of a group of structurally very similar potent neurotoxins isolated from the marine zoanthid *Palythoa toxica* or other *Palythoa* spp. Each consists of a single carbon chain of 115 carbon atoms containing 64 chiral centers and seven double bonds, and terminating in nitrogen-containing groups; there are 128 carbons in all and a further double bond in a methylene group. The chain bears numerous hydroxyl groups and a number of methyl groups, and in several places it is folded and formed into pyran rings.

Pancreas

A compound gland, occurring in the abdominal cavity of most vertebrates, that has both exocrine and endocrine functions. The major (exocrine) part consists of acinar tissue, which secretes pancreatic juice into the upper part of the gut (duodenum in mammals). Within this acinar tissue are scattered numerous (endocrine) islets of Langerhans, containing notably A cells, secreting glucagon, B cells, secreting insulin, and D cells, secreting somatostatin. —pancreatic adj.

Pancreastatin

a 49-residue peptide hormone that inhibits glucoseinduced insulin release and exocrine pancreatic secretion. It was originally isolated from porcine pancreas, where it is colocalized with insulin, glucagon and somatostatin, but it is part of the sequence of chromogranin A and is thus found in tissues where the chromogranin gene is expressed, including most endocrine cells

Pancreatic Juice

A slightly alkaline digestive juice secreted by the exocrine pancreas into the upper part of the small intestine. It contains numerous enzymes and inactive enzyme precursors including a-amylase, chymotrypsinogen, lipase, procarboxypeptidase, proelastase, prophospholipase A2 , ribonuclease, and trypsinogen. Its high concentration of hydrogen carbonate ions helps to neutralize the acid digesta from the stomach.

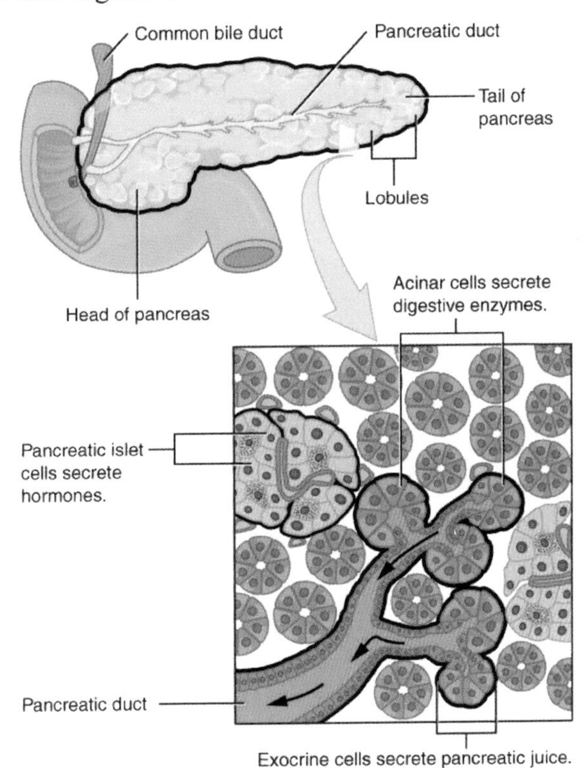

Figure 33. Exocrine and Endocrine Pancreas.

Source: Image by Wikimedia Commons

Pancreatic Lipase

A Ca2+-requiring triacylglycerol lipase (EC 3.1.1.3) that is secreted into the intestine by the exocrine pancreas when stimulated by cholecystokinin in response to ingestion of food. It degrades triacylglycerols, partially or completely, to fatty acids and glycerol in the intestine; it acts at an ester–water interface. It is a glycoprotein with sequence similarity with other lipases.

Pancreatic Thread Protein or Pancreatic Stone Protein

(abbr.: PSP); a C-type lectin that may inhibit spontaneous calcium carbonate precipitation, and one of the major secretory proteins of the human exocrine pancreas. It is found in acinar cells of pancreas and (in smaller amounts) in the brain. It is a major soluble protein of human pancreatic calculi, and a Ca2+- binding phosphoprotein present in zymogen granules of pancreatic acinar cells, and secreted in the pancreatic juice of normal subjects and calculus formers.

Pancreozymin

abbr.: PZ; a hormone having secretagogic activity on the exocrine pancreas and extractable from duodenal mucosa. It is identical to cholecystokinin.

Panning (in cell biology)

A method for enriching a population of cells, or phages, displaying peptides or fragments of antibodies by allowing them to bind to ligands immobilized on a solid surface such as that of a microtitre plate. Washing the wells of the plate removes nonadherent or weakly adherent cells or phages, leaving those adhering as a result of stronger interactions, to be collected for further study. Several cycles of enrichment are used.

Pantetheine

N-pantothenylcysteamine; the D-enantiomer is a growth factor for *Lactobacillus bulgaricus*, and is an intermediate in the pathway for the biosynthesis of coenzyme A in mammalian liver and some microorganisms.

Pantothenate Kinase

abbr.: PANK. EC 2.7.1.33; a kinase enzyme that is specific for phosphorylation of pantothenate to 4′-phosphopantothenate, being the first reaction in the biosynthesis of coenzyme A. Many mutations in the gene for PANK2 (locus at 20p13) produce enzyme deficiency and Hallervorden–Spatz syndrome, a neurodegenerative disease involving the accumulation of iron in the brain.

Papain or Papaya Peptidase I

EC 3.4.22.2; a cysteine endopeptidase obtained from the latex, leaves, and unripe fruit of the papaya (or pawpaw) tree, Carica papaya. It will preferentially hydrolyze peptide bonds at the carbonyl end of Arg, Lys, and Phe residues (but never Val), with a preference for large hydrophobic residues at the P2 position. It also has esterase, thiolesterase, transamidase, and transesterase activity. Papain is unusually stable to elevated temperatures and to denaturing agents and consists of a single polypeptide chain of 212 amino-acid residues.

Papaverine

1-[(3,4-dimethoxyphenyl)methyl]-6,7-dimethoxyisoquinoline; a constituent of opium that acts as a smooth muscle relaxant; this action is thought to be due to phosphodiesterase inhibitory activity and blockade of membrane calcium channels. Like codeine and morphine, it is a metabolic derivative of (S)-reticuline.

Paper Chromatography

abbr.: PC; a technique of chromatography, applicable to microgram quantities of soluble substances, in which specially prepared filter paper, chromatography paper, forms the support for the stationary phase. The latter is commonly a film of water held by adsorption on the cellulose fibers of the paper and in equilibrium with a water-immiscible liquid or liquid mixture, which forms the mobile phase. A solution of a sample is applied near one end of a strip of paper, allowed to dry, and the mobile phase allowed to flow over the strip from that end.

Papovaviridae

a family of DNA viruses, most or all of which are, under suitable conditions, oncogenic in vertebrate hosts. The virion, 45–55 nm in diameter, consists of a nonenveloped icosahedral capsid with 72 capsomeres and contains a circular DNA genome.

Papovavirus

Any virus belonging to the family Papovaviridae. The name derives from papilloma, polyoma, and vacuolating agent, an early name for simian virus 40.

PAPP-A

abbr. for pregnancy-associated plasma protein A; a plasma protein first found in the serum of pregnant women but also present in men. It is a zinc metalloprotease secreted by fibroblasts and cleaves insulin-like growth factor binding protein (IGFBP)-4 to release insulin-like growth factor (IGF).

PAR

abbr. for protease-activated G-protein-coupled receptor; any G-protein-coupled membrane receptor of platelets and endothelial cells that becomes activated on binding thrombin and cleavage of the extracellular segment of the receptor. Most mouse embryos that lack PAR1 die in mid-gestation.

Parallel

1 (in biochemistry) describing a pair of linear structures, such as two polynucleotide or polypeptide chains, that are polarized or asymmetric in the same direction. 2 (in physical chemistry) denoting the spins of a pair of electrons, occupying the same atomic or molecular orbital, that are described by the same spin quantum number. Compare antiparallel. 3 (in electricity) describing an electrical component that is connected to the same two points in a circuit as another component; describing two or more components that are so connected.

Paralogue or (esp. US)

paralog is a gene, protein, or biopolymeric sequence that is evolutionarily related to another by descent from a common ancestor, having diverged as a result of a gene duplication event within an organism. Paralogues usually perform different but related functions within that organism, for instance, human red and green opsins. –paralogous adj.

Paramagnetism

The property displayed by substances that have a positive but small magnetic susceptibility, due to the presence in them of atoms with permanent magnetic dipoles caused by unpaired electron spins (with a contribution from the orbital motion of the electrons). These dipoles tend to align themselves in the direction of an applied magnetic field but no permanent magnetism is conferred on such substances.

Parameter

1 (In mathematics) an unknown quantity that is a constant in a particular context but that may have different values in other (similar) contexts; e.g. the coefficient b in the general equation, $y = bx + c$, representing a family of straight lines. 2 (in statistics) a numerical characteristic of a population (as opposed to that of a sample of such a population). 3 any distinguishing characteristic of something, especially one to which a measured value is or can be ascribed. —parametric adj.

Paraprotein

Name originally given to any plasma protein that gave an abnormal band on electrophoresis, derived from para+ + protein. Usually, such a protein is a monoclonal immunoglobulin derived from neoplastic plasma cells, and present at abnormally high concentrations in the blood plasma. Such proteins are seen as a discrete band in the gamma-globulin region, but they may appear elsewhere if the paraprotein is IgA or IgM. Examples are proteins characteristic of a myeloma, e.g. Bence-Jones protein, amyloid proteins, Waldenstrom's macroglobulinemia, or cryoglobulins. Some paraproteins are not abnormal, e.g. the immunoglobulin that arises as a result of a severe bacterial infection.

Paraquat

methyl viologen; 1,1′-dimethyl-4,4′-bipyridynium dichloride; a nonselective herbicide whose action requires direct contact with the plant. It is highly toxic to humans and other animals if ingested in undiluted form.

Paratartaric Acid

An old name for racemic DL-tartaric acid. The resolution of this acid into the D- and L-enantiomers by Pasteur was a landmark experiment in stereochemistry.

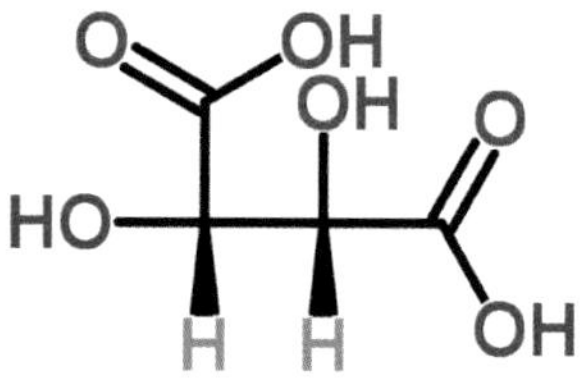

Figure 34. Structure of Tartaric Acid.

Source: Image by Wikimedia Commons

Parathyroid Hormone-Related Protein

abbr.: PHRP; a protein (141 amino acids) that is homologous in the N-terminal region with parathyroid hormone, has effects similar to this, and binds the same receptor. It is present in breast and milk, kidney, brain, and other tissues, especially during intrauterine life, and seems to function largely in a paracrine fashion.

Paratose

3,6-dideoxy-D-glucose; 3,6-dideoxy-D-ribo-hexose; a monosaccharide that occurs in type A O-antigen chains of the lipopolysaccharide on the outer membrane of certain species of Salmonella.

Parkinson's Disease or Parkinsonism

A progressive disorder of the central nervous system characterized by tremor and impaired muscular coordination thought to be due to defective dopaminergic transmission in some parts of the brain. There is a loss of dopaminergic neurons connecting the substantia nigra with the striatum, resulting in loss of dopaminergic inhibition in the striatum, with resulting cholinergic hyperactivity.

Parotid Gland

Either of a pair of salivary glands situated near each ear in mammals. The duct runs forwards and empties into the oral cavity.

PARP-1

abbr. for poly(ADP-ribose) polymerase 1; in mammals, a highly conserved protein (113 kDa) that is strongly activated by DNA strand breaks, and participates in modulating DNA base excision repair, apoptosis, and necrosis. It includes a DNA-binding domain with 23 zinc finger motifs, and a poly(ADP-ribose) polymerase catalytic domain. Mouse gene knockouts for this protein are remarkably resistant to myocardial infarction, stroke, shock, diabetes mellitus, and neurodegeneration.

Pasteur Effect

Or (formerly) Pasteur–Meyerhof effect either the phenomenon that occurs in facultatively anaerobic cells whereby oxygen inhibits glycolysis or fermentation, or its converse whereby the rate of glycolysis or fermentation increases when oxygen is excluded.

Q

Quad Denoting

Quar denoting can be defined as four dynamically significant products or substances in an enzyme mechanism.

Quadrupole Mass Spectrometer

Quadrupole mass spectrometer is a sort of mass spectrometer in which the molecular ions are isolated by a mass filter containing four metal rods with an alternating high-frequency voltage (quadrupole field).

Quality Factor Or (Formerly) Relative Biological Effectiveness

Quality factor or formerly known as relative biological effectiveness is an index of the capability of a given sort of ionizing radiation to cause biological damage, contingent upon the density of the ionization produced comparatively with that created by gamma radiation. For all beta particles (+ or –) and X-(and gamma) radiation prone to be experienced in the utilization of radioisotopes it is assigned a value of unity, for all alpha particles a value of 10 and for thermal neutrons usually a value of 3.

Quantitative Trait Locus

QTL is abbreviated for quantitative trait locus. It is a region of a chromosome containing genes that are presumed to make a critical contribution to the expression of a complex phenotypic trait. Quantitative traits (QTs) are usually impacted by the environment and by more than one gene and incorporate numerous medically notable traits (for example obesity, blood pressure, etc.). Specific genes can be isolated by other molecular methods, once mapped to little chromosomal districts, for example utilizing linkage maps.

Quantity Or Physical Quantity

In systems of measurement, any property whose value can be denoted as the product of a unit and a numerical value. There are seven base quantities in the International System of Units (see SI), physical quantities considered as being independent dimensionally. These are: luminous intensity, length, amount of substance, mass, thermodynamic temperature, time, electric flow, thermodynamic temperature, the measure of substance, and glowing power; all are randomly characterized. Derived units are communicated arithmetically in terms of the base units. Physical quantities symbols are printed in italics type and they are generally single Greek or Roman letters and might be capital or lower case.

Quantize Or Quantise

To limit a physical quantity to a set of discrete values, described by quantum numbers. A quantized physical quantity can't vary consistently but should change in steps.

Quantum (Pl. Quanta)

1. a specific amount or quantity of something.
2. a unit amount of a physical quantity, mainly of electromagnetic energy, to such an extent that all other amounts are integral multiples of it.

A quantum of energy taken up during radiation absorption or released during radiation emission is equal to h m, where h is the Planck constant and m is the frequency of the radiation in hertz according to the quantum theory. A quantum of light or other electromagnetic energy is described as a photon.

Quantum Mechanics

Quantum mechanics is a mathematical theory created from the quantum theory (see quantum) as a substitution for classical mechanics to describe satisfactorily the behavior of elementary particles, atoms, and molecules in terms of observable quantities like the frequencies and intensities of spectral lines.

Quantum Number

A Quantum number is a set of half-integral or small integral numbers that give the numerous potential values of a quantized property of a system. Quantum numbers are utilized particularly in depicting the properties of elementary particles, like their charge (values of -1, 0, or +1) or spin (upsides of -1/2, or +1/2), and in determining the energy states of electrons in atomic orbitals, every one of which is described by a unique set of four quantum numbers.

Quantum Requirement

In photosynthesis, the quantum requirement is the inverse of quantum yield (def. 2), for example, the number of quanta of light energy needed to achieve the release of one molecule of dioxygen. The theoretical value is 8, however, the value noticed in an experimental system is by and large to some extent higher, contingent upon conditions.

Quantum Yield

Quantum yield is abbreviated as Q. In a luminescent system the likelihood of luminescence happening in given conditions is demonstrated by the ratio of the number of photons, for example, quanta of light, emitted by the luminescing species to the number absorbed. To compare quantum efficiency, The measurement of Q requires the counting of photons, in light of the fact that:

Q = photons emitted/photons absorbed

It requires exceptionally specialized instrumentation, yet a less conscientious method might be utilized in which the fluorescence of a fluor is contrasted and that of a fluor of known Q. Then, at that point,

$$Q_x = [I_x Q_s A_s]/[I_s A_x]$$

where Qs is the quantum yield of the standard, I_x and I_s are the fluorescence intensities of the standard and sample, and A_x and A_s are the percentage of absorption of each solution at the exciting wavelength.

In photosynthesis, the inverse of quantum requirement; the fractional number of molecules of carbon dioxide reduced per photon absorbed.

Quartz

Quartz is a glassy form of silica, often transparent, a colorless, SiO_2 utilized in optical devices and cuvettes as a result of its transparency to near-ultraviolet radiation. It is enantiomorphous, the left and right-handed forms rotating the plane of polarized light in opposite directions.

Figure 35. An Illustration of Quartz.

Source: Image by Wikimedia Commons

Quasi-Species

In virology, quasi-species is the number of inhabitants of closely related however non-identical viral genomes that evolve over time through spontaneous mutations within an individual infected with a single genotype. Diversification through the emergence of quasi-species is one mechanism by which viruses evade the host immune response.

Quaternary Ammonium Compound

Quaternary ammonium compound is a compound that can be considered as derived from ammonium salt or an ammonium hydroxide by substitution of every one of the four hydrogen atoms of the NH_4^+ ion by organic groups. Certain compounds of this class, ones in which one of the organic groups is a long-chain (C8-C18) alkyl group and the other three are shorter-chain alkyl or other groups have the properties of cationic detergents and are strong antimicrobial agents. They are bactericidal at higher concentrations and bacteriostatic at low concentrations, being by and large more active against Gram-positive than Gram-negative organisms. They show relatively low toxicity to higher humans and animals and subsequently are broadly utilized as disinfectants and antiseptics.

Quaternary Structure

Quaternary structure is the fourth-order of complexity of structural organization displayed by nucleoprotein molecules, protein, and nucleic acid. It alludes to the arrangement in space of the subunits of a multimeric macromolecule, and the ensemble of its intersubunit interactions and contacts, regardless of the internal geometry of the subunits. Subsequently, it is possessed only when the molecule in question is made up of at least two (non-identical or identical) subunits that are at least potentially separable, i.e. are not linked by covalent bonds.

Quellung Reaction

Quellung reaction or Neufeld Quellung reaction or pneumococcus capsule swelling reaction the expanding of the capsule is noticed when the particular antibody is blended with suitable bacterial cells, for example, pneumococci. It is presumably because of the deposition of antibodies on the outside of the capsule, making the latter visible clearly. [From the German, Quellung, enlarging; depicted by Ferdinand Neufeld (1869-1945), German bacteriologist, in 1902.]

Quenching

1. Quenching is the destruction or reduction of luminescence, particularly fluorescence, of a sample by the inclusion or addition of a quencher. Fluorescence quenching is the basis of a practical technique for studying the binding of small molecules to proteins or different macromolecules.
2. the decrease by all means possible for the efficiency of energy transfer from beta particles to the photomultiplier(s) in a liquid-scintillation counter.

Such means include:

1) dilution quenching – reduction of the probability of scintillation occurring through dilution of the scintillator by the sample;
2) chemical (or impurity) quenching – absorption by a component of the sample of a portion of the energy of the beta particles without producing photons, or without fluorescing and in competition with the essential fluor of a portion of the photons transmitted by emitted by excited solvent molecules; and
3) color quenching - absorption by a colored component of the sample of some of the photons emitted by the secondary fluor.

1. the termination of an enzymic or chemical reaction by a sudden change of the conditions or addition of another reagent.

Quercetin

3,3′,4′,5,7- pentahydroxyflavone; a member of the flavonoids, or flavones, widely distributed in plants, usually as glycosides; it is the aglycon of rutin, quercitrin, and various other materials. Known in folk medicine for its anti-inflammatory properties, it hinders the lipoxygenase pathway, has some antiplatelet and anti-thrombotic activity in vivo, inhibits mast-cell degranulation, and possesses anti-asthmatic activity. It inhibits numerous enzymes, including protein kinases, and furthermore inhibits protein synthesis, DNA, and RNA. Some of its actions might reflect its property as a nucleoside antagonist.

Quin - 2

abbr. for 2 - [(2 – amino – 5 - methylphenoxy) methyl] – 6 – methoxy 8 – aminoquinoline - N,N,N′,N′ - tetraacetic acid; a molecule functionally related to the Ca^{2+} - chelating agent EGTA, but holding fluorescent groups.

The quantum yield of fluorescence is substantially enhanced on binding Ca^{2+}, leading to its use as a quantitative indicator of Ca^{2+} concentration. It is generally utilized as its tetraacetoxy ester, which readily crosses the plasma membrane and is then hydrolyzed within the cell by esterases to Quin-2, the polar nature of which causes it to be retained as an intracellular indicator.

Quinacrine Or Mepacrine OrAtebrine

6 – chloro – 9 - (4 – diethylamino – 1 - methylbutylamino) – 2 - methoxyacridine; a derivative of acridine, formerly utilized as an anthelmintic and antimalarial. It can be utilized as a fluorochrome to label DNA in chromosomes and fluoresces powerfully in UV light.

Quinhydrone

Quinhydrone is an addition compound of one mole each of p-benzoquinone and hydroquinone. It is a reddish-brown crystalline substance with a dark-green luster, useful as a redox system in the quinhydrone electrode and as an antioxidant in photography.

Quinhydrone Electrode

Quinhydrone electrode is a half-cell comprising of bright platinum or sometimes gold electrode submerged in a test solution to which a little quinhydrone has been added. It is useful as a secondary hydrogen electrode for the electrometric measurement of pH since it has an electric potential related to the pH value of the solution.

Quinine

6′-methoxycinchonan-9-old; an alkaloid which is bitter-tasting acquired particularly from the bark of cinchona, any of various shrubs or trees of the tropical genus Cinchona (particularly the Javan species *C. lidgeriana* and different South American species). It was previously much utilized and sometimes is still required for the treatment and prevention of malaria. It attaches to the DNA of the malarial parasite *Plasmodium* spp., and is thought consequently to hinder the biosynthesis of the parasite's nucleic acids. Quinine and beverages containing it help to prevent cramps, for example, tonic water.

4 - Quinolone Antibiotic

Quinolone antibiotic is a generic term for any of a wide range of synthetic antibacterial compounds whose molecular structures contain a 4-oxo-1, 4-dihydroquinoline, or a 4-oxo-1, 4-dihydrocinnoline nucleus; for example oxolinic acid, cinoxacin, ciprofloxacin, nalidixic corrosive, and ciprofloxacin. Their antibiotic activity obtains from interplay with the A subunit of bacterial DNA gyrase (see type II DNA topoisomerase); this outcome in the trapping of bacterial DNA in the complex forms with the A subunit. Rejoining of the DNA strands broken by the enzyme is in this way captured.

Quinone

1. Quinone is a popular name for p-benzoquinone.
2. any member from a class of diketones derivable from aromatic compounds by conversion of two CH groups into CO groups with any essential rearrangement of double bonds. Simple quinones are typically p-quinones, that is, derivatives of p-benzoquinone (compare hydroquinone (def. 2)), however they may likewise be o-quinones, that is derivatives of o-benzoquinone (compare catechol (def. 2)). Naturally occurring quinones form a varied, large, and widespread group of compounds. They incorporate various electron carriers, numerous pigments, for example, ubiquinone, plastoquinone, and vitamin K.

Quinoprotein

Another name for quinoprotein is quinoenzyme. any enzyme protein that contains a quinone cofactor, that is the freely dissociable pyrroloquinoline quinone (PQQ) or methoxantin, or the covalent species topaquinone (TPQ), tryptophan tryptophylquinone (TTQ), or lysine tyrosylquinone (LTQ) that are inferred by post-translational change of few proteins. PQQ is a characteristic for oxidoreductase enzymes within the sub-subclass EC 1.1.99, which incorporates alcohol dehydrogenase (acceptor) (EC 1.1.99.8) of methylotrophic bacteria, glucose dehydrogenase (pyrroloquinoline-quinone) (EC 1.1.99.17), and various dehydrogenases acting on other specified (kinds of) polyols or alcohols. TTQ occurs in bacterial amine dehydrogenase, TPQ occurs in bovine serum copper amine oxidase, and LTQ occurs in bovine aorta lysyl oxidase (EC 1.4.3.13).

Quinovose

The symbol for quinovose is Qui. it is a trivial name for 6-deoxy-D-glucose. A quinovose occurs, for example, as its 6-sulfo derivative (6-deoxy-a-D-glucopyranosyl 6-C-sulfate) in the sulfoquinovosyl diacylglycerols, which structure one of the two significant groups of the glycerolipids of chloroplasts. The sulfoquinovosyl derivative likewise occurs in a few cardiac glycosides.

R

R Symbol For

1. a residue of the a-amino acid L-arginine (alternative to Arg).
2. a residue of an incompletely specified base in a nucleic acid sequence that may be either guanine or adenine.
3. a residue of an undefined purine nucleoside (alternative to Puo).
4. an undefined univalent group in a formula of an organic compound.
5. The group should be appended through carbon and derived from a heterocyclic, aliphatic, or carbocyclic compound. When different, up to three such groups, might be designated R, R, and R, or R1, R2, and R3; superscript numerals only should be used for more than three such groups.
6. röntgen (a non-SI unit of radiation openness).
7. resonance effect (in electron displacement).

RAB

A quality encoding a Rab protein - of a group of GTP-restricting proteins like Ras proteins (see RAS) and presumably associated with layer traffic, and so on Rab proteins contain dicysteine themes close to their C ends to get a geranylgeranylyl moiety.

Rabbit Aorta-Contracting Substance

Abbreviated as RCS is a name given (in 1969) to an unsteady component set free from guinea-pig lung that caused compression of segregated pieces of rabbit aorta. The action was most likely because of the presence of thromboxane A_2.

Rabbit Reticulocyte System

A cell-free system, from lysed rabbit reticulocytes, that deciphers added eukaryote mRNAs from a wide assortment of heterologous sources. Endogenous mRNA is degraded utilizing a Ca2+-subordinate ribonuclease, micrococcal endonuclease; this catalyst is inactivated by adding EGTA. Heterologous mRNA is then added and is deciphered by the system.

RACE

Abbreviated for fast intensification of cDNA closes is a minor departure from the polymerase chain response intended for enhancement of succession relating with the 5′ or 3′ finishes of specific mRNAs. The previous is significant for the area of the record start site as well as giving full-length cDNA successions.

Racemase

A chemical that interconverts the two enantiomers of a chiral compound; such proteins are classified in subclass EC 5.1 (which likewise incorporates epimerases). On the off chance that more than one chiral center is available, a racemase should alter the arrangement of all the chiral centers. Hence, the name methylmalonyl-CoA racemase, once in a while utilized for methylmalonyl-CoA epimerase (EC 5.1.99.1), is erroneous - the compound doesn't change designs of the chiral centers of the CoA moiety.

Racemic

Image: (±)- or (previously) dl-; It indicates the presence of equimolar measures of the dextrorotatory and laevorotatory enantiomers of a compound, whether or not the (optically inactive) item formed is glasslike, fluid, or vaporous. A homogeneous strong stage made out of equimolar measures of enantiomeric atoms is named a racemic compound, and a combination of equimolar measures of enantiomeric particles present as individual stages is named a racemic mixture; see likewise racemate. A (±)- item might be settled into its (+)- and (-)- parts. The term is obtained from racemic acid [from Latin racemus, pack of grapes], the optically inert combination of (+)- and (-)- tartaric acids once in a while found during the assembling of wine. Likewise, the term is applied to DL-and (RS)- blends. See D/L show, optical isomerism, grouping rule.

Rad 1

The unit for radian. 2 image: rad or (to stay away from rad (def. 1)) rd; the cgs unit of absorbed portion of ionizing radiation, comparable to a retention of 0.01 joule of energy per kilogram (= 100 ergs for every gram) of lighted material. Though the rad addresses, by and by, about a similar measure of energy as the röntgen (the specific comparability relying upon the material lighted), it applies to an ionizing radiation in any medium. 1 rad = 0.01 Gy. 3 abbr. for radiation.

Radiation

1. The emanation of energy from a source and its transmission as particles, beams, or waves, particularly as electromagnetic beams or waves, sound waves, or surges of subatomic particles.
2. Energy so transmitted, especially the corpuscular and electromagnetic rays emitted in the decay of a radioactive nuclide.
3. Ionizing radiation.
4. Parting from a common point, especially radially from a central point or source.

Radiation-Chemical Reaction

Any compound response that is started by the assimilation of radiation however is visible from a photochemical response by its lack of accuracy (for example by leading to an assortment of response items) and by happening at higher radiation energies (for example by ingestion both of electromagnetic radiation from the mid-bright area onwards or of high-energy particulate radiation) and hence by coming about consistently in the development of visible particles.

Radiation Hybrid Map

Abbreviation as RH map, it's a guide of lighted chromosomal destinations got from cross-breed cells containing pieces of illuminated chromosomes. RH maps give significant markers, permitting the development of exact grouping labeled site (STS) maps (with the STSs situated by their recurrence of partition by radiation-instigated breaks). These are helpful in requesting genetic loci along chromosomes and in studying multifactorial diseases.

Radiation Inactivation Method

A strategy, in view of a target hypothesis, that empowers the size of the functional unit of a naturally dynamic macromolecule (or of its structures with different particles) to be assessed; it tends to be utilized on unpurified (for example membrane bound) material. Assessments are made of the organic action staying in a sample after exposure to different huge dosages of ionizing radiation (electrons or X-or gamma radiation of somewhere around 1 MeV), and the worth, D37, of the radiation portion, in Mrad, expected to decrease the action to 37% of (for example e-multiple times) its valuuue is gotten; the overall atomic mass, Mr , of such a macromolecule may then be found from the experimental relationship Mr = 6.4 × 105/D37, where D37 set in at 30°C.

Radical

1. An atomic element, charged or uncharged, that has an unpaired electron (however ordinarily barring any paramagnetic metal particle); regularly framed by homolysis of a covalent bond. A radical is shown in an equation by a dot representing the unpaired electron and put (if conceivable) alongside the image; for example, HOO•, •CH3. The term radical is currently ideally limited to any radical that doesn't shape part of apair.
2. A previous name for group.

Radical Pair

1. Any two radicals (def. 1) in nearness in fluid arrangement, with a oond. They might have been formed all the while, for example by homolysis, or have met up by dissemination. While together, the relationship of their unpaired electron happens and shows itself as bending of magnetic reverberation spectra.
2. A cation and anion radical formed momentarily in the essential photochemical course of photosynthesis.

Radical Scavenger

Is a substance that can respond promptly with, and consequently target, radicals. In natural systems or materials, scavengers might work as cell supports or may shield from harm by ionizing radiation. Among the significant radical scavengers that capacity in tissues at physiological levels are the water-dissolvable substances ascorbate, glutathione, and the purine bases, and the fat-solvent substances tocopherols, the retinols, the carotenes, and the ubiquinones.

Radioactivate

To change over a steady nucleus into one going through radioactive decay by high velocity particles, including protons, alpha particles, neutrons, and deuterons. These are formed in a cyclotron or synchrotron from a radioactive source, and sped up through a potential differenceof two or three thousand volts by a magnetic field.

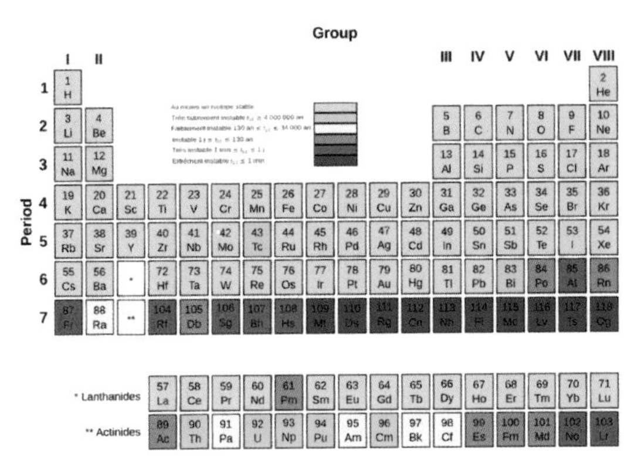

Figure 36. Periodic Table Radioactivity.

Source: Image by Wikimedia Commons

Radioactivation Analysis or Activation Analysis

Is a technique for the subjective as well as quantitative assessment of the compound components in an example? It relies upon the distinguishing proof and assurance of the radionuclides framed when the example is assaulted with neutrons or different particles.

Radioactive Isotope

Any isotope of a substance component whose nucleus is unsteady and emanates alpha, beta, or gamma rays. The item is another component, which might be steady or unsound; assuming that the last option, it decays further. Hence, the vital isotope of uranium,^{238}U, transmits an alpha molecule and decays to ^{234}Pa, which emanates a beta molecule forming ^{234}Th; this chain proceeds to the development of ^{206}Pb, which is steady.

Radioactive Tracer

A radioactive substance which is added to a metabolic system in amounts (mass) too little to damaging the system, to follow, by isolating the radioactive intermediates formed, the grouping of changes gone through regularly by the equivalent (nonradioactive) atom under comparative circumstances.

Radioallergosorbent Test

Abbreviate as RAST Is a test utilized principally for measuring the degrees of antigen-specific immunoglobulin E (IgE) in serum. A particular allergen, which has been covalently coupled to an insoluble Sephadex transporter, is incubated with serum containing the test measure of IgE. The transporters are then washed and further incubated with radiolabelled antibodies to IgE, how much radioactivity bound being a proportion of how much IgE. The upside of this technique over the radioimmunosorbent test is that IgE specific to a specific antigen can be specifically measured.

Radiofrequency Radiation

The electromagnetic radiation incorporating frequencies in the range 1 mm (300 GHz) to 30 km (10 kHz) (frequencies in brackets). This range incorporates microwave radiation and lies over that of infrared radiation.

Radiography

The utilization of X-rays to analyze the inward designs of a body. An X-ray film is set behind the body, through which X-rays are passed. After exposure, there are areas of the film in which improvement has not happened because of retention of the X-rays the body, in this way yielding an example of these structures.

Radioimmunoassay

Abbreviated as RIA Is an exceptionally sensitive strategy for the measure of nonradioactive material. A known measure of immune response (Ab) coordinated against the substance (antigen, Ag) to be examined is incubated with a combination of Ag and radioactive Ag (Ag*), so that absolute (Ag + Ag*) is in abundance. At the point when nonradioactive Ag is added to Ab alongside Ag*, Ag and Ag* compete for restricting to Ab, so that less Ag* will be found in the complex as the proportion Ag:Ag* increments. On the off chance that the immune response antigen complex is, isolated from free antigen, the aggregate sum of antigen (the substance being measured) can be determined from the proportion Ab-Ag*: Ag* and the known titer of counter acting agent. It is regularly helpful to bind the antigen to plates with the goal that how much ligand bound to the plate is relative to how much test antibodies are present.

Radio - Immunosorbent Test

Abbreviated as RIST is a test utilized fundamentally for measuring whole-serum immunoglobulin E (IgE) levels. Antibodies specific for IgE, covalently connected to an insoluble dextran, are incubated first with a known measure of radiolabelled IgE and afterward with the serum containing the test measure of IgE. From the lessening in how much radiolabel bound to the immobilized antibodies, because of competition by the unlabelled IgE in the serum, the aggregate sum of IgE in the serum can be assessed.

Radiometer

1. An instrument used to gauge bright energy.
2. Crookes radiometer is a particular sort of radiometer in which the different conduct of a cleaned surface, which reflects energy, and a darkened surface, which ingests it, is used to make a rotor turn at a rate connected with the power of radiation.

Figure 37. An Illustration of Radiometer.

Source: Image by Wikimedia Commons

Radiomimetic

1. Depicting any medication or other synthetic substance, commonly an alkylating specialist, whose impacts in living systems (for example cancer-causing, immunosuppressive, mutagenic) look like those delivered by ionizing radiation; all the more freely, portraying any such specialist ready to cause quality and chromosome changes.
2. A radiomimetic (def. 1) substance or specialist.

Radixin

It is a protein of the erythrocyte band 4.1 family present in many eukaryotic cells; it has around 75% similarity to human ezrin. It is an actin protein, profoundly gathered in the the cell to-cell adherens intersection and the cleavage furrow in the interphase and mitotic stage.

RAF

Any quality encoding a Raf protein, of which there are no less than three: Raf-1, A-Raf, and B-Raf. These are protooncogene cytoplasmic serine/threonine protein kinases (EC 2.7.1.-) of the Ras/Mos family; they are homologous with members from the protein kinase C family in both their C-terminal and N-terminal groups. Human Raf-1 has binding spaces for ATP, phorbol ester, and diacylglycerol. This protein is engaged with the transduction of mitogenic signals from the cell film to the nucleus. In the mouse, raf is the changing quality of the mouse sarcoma infection, 3611, v-Raf has a shortened operator that causes constitutive actuation of kinase functions. The c-Raf-1 protein kinase, a 74 kDa protein, has been found in numerous mammalian cell types. Raf-1 is vital for cell development, change, and separation and is initiated by a wide assortment of extracellular signals. It has MAP kinase activity. In numerous cell types, Raf works quickly downstream of the G-protein Ras in the initiation pathway; Ras intervenes movement of Raf-1 to the plasma layer, where it is activated. In certain cells Raf might be phosphorylated by a protein kinase C. The physiological substrate for Raf kinase is a protein kinase (MEK) of yeast that is expected for meiotic recombination and which on phosphorylation can initiate MAP kinase.

Raffinose

Gossypose or melitose or melitriose the trisaccharide O-aD-galactopyranosyl-(1→6)- O-a-D-glucopyranosyl-(1→2)- b-D fructofuranoside; it happens in plants as generally as sucrose, being available, e.g., in cereal grains, cotton seeds, and numerous vegetables. It is the principal member from a series wherein galactosyl deposits are connected to sucrose, others being the tetrasaccharide stachyose and the pentasaccharide verbascose in which a-D-galactopyranoside of myo-inositol (for example galactinol) is the giver of the galactosyl residues.

Raft

A sphingolipid-and cholesterol-rich microdomain in a plasma layer. Rafts are equal in structure to caveolae, however, need caveolin, and are available in all cell types. They isolate and think GPI-anchored, doubly acylated membrane proteins and transmembrane proteins. They might be forerunners of caveolae.

Ragged-Red Fibers

Vigorous skeletal muscle strands that amass broadened and unusual mitochondria containing abnormalities when they are ischemic or anoxic. The mitochondrial systems seem red on staining with Gomori adjusted Trichrome. They are generally connected with mitochondrial diseases.

Raman Scattering

The random scattering of light when it is reflected from solids, liquids or gases molecules. The Raman effect is comprised of apprearane of certain spectral lines near the point of incident of light. The Raman lines appear light and weaker as compared to the original wavelength of the light. The Raman scattering is of two types, i.e., Stokes scattering and anti-stokes scattering depending on the frequency of emitted radiation as compared to the original incident radiation.

Rayleigh Scattering

The versatile scattering of photons of light by particles or molecules of the substance through which the light passes. At the point when the light goes through a fluid or gas, particles or particles are energized by the electric vector of the light, consequently prompting a quickly fluctuating dipole in the atoms or molecules in the light way. The fluctuating dipole prompts the discharge of electromagnetic waves in different orientations of a similar recurrence as the incident radiation, and this emanation is viewed as scattered light.

RBS

Abbreviate for RénéBorghgraef arrangement; Is the name for universally useful surfactants and disinfecting specialists, particularly for research facility use, that have joined cleanser and chelating properties. They are related to a number or other assignments, for example, RBS 25, RBS 35, RBS strong.

Reactive Oxygen Species

Abbreviating as ROS- Is an exceptionally dynamic oxygen animal categories, for example, superoxide particle, hydroxyl radical, hydrogen peroxide (H_2O_2), or peroxynitrite. Most ROS are created as poisonous results of oxidative phosphorylation in mitochondria. H_2O_2 is likewise created by different responses in peroxisomes, and by phagocyte NADPH oxidase, thyroperoxidase, and xanthine oxidase in specific cell types. Cells safeguard themselves through the superoxide dismutase, glutathione peroxidase, and catalase activities, and the cancer prevention agents (for example glutathione, carotene) they contain. Ongoing exposure to ROS makes oxidative harm mitochondrial and cell proteins, lipids, and nucleic acids.

Readthrough Protein

A molecule formed by the interpretation of a messenger RNA when one of its end codons is misread, or is perceived by a silencer tRNA. Polypeptide-chain union then, at that point, go on until the following stop codon is reached, bringing about the formation of a variation protein with a C-terminal expansion.

Receptor

Any cell macromolecule that ties a chemical, synapse, drug, other agonist, or intracellular messenger to start an adjustment of cell work. Receptors are concerned directly and specificly in substance motioning between and inside cells. Cell surface receptors, for example, the acetylcholine receptor and the insulin receptor, are situated in the plasma layer, with their ligand-restricting site presented to the outer medium. Receptors may likewise be available in layers of the sarcoplasmic reticulum (for example ryanodine receptor), T tubule (for example dihydropyridine receptor), or endoplasmic reticulum (for example inositol trisphosphate, or IP3 receptor). Intracellular receptors (for example steroid-chemical receptors) tie ligands that enter the cell across the plasma layer. These can be situated in the cytosol or nucleus. Receptors can be synergist, catalyst related, or G-protein-related.

Receptor Activity Modifying Protein

Abbreviating as RAMP Is a single transmembrane-domain protein expected for articulation of calcitonin gene associated peptide or adrenomedullin receptor aggregates from the calcitonin receptor-like receptor quality (CRLR). For full activity, CRLR structures heteromers with RAMP and receptor part protein.

Recombinant DNA

A piece of deoxyribonucleic acid (DNA) that has been embedded into a cloning vector, in this manner prompting its utilization in the release of a clone of cells described by the presence of the gene. The term is obtained from the idea that addition into the vector is a type of hereditary recombination. [Note: The abbreviation rDNA has some of the time been applied to recombinant DNA, yet this utilization is deterred (rDNA having been pre-empted for ribosomal DNA); while choices, for example, recDNA or rtDNA have been recommended, IUBMB considers a standard abbreviation unnecessary.]

Recombinant DNA Technology

An area of biotechnology concerned about the control of recombinant DNA. It has numerous significant applications, including

1. DNA sequencing, which might prompt the capacity to anticipate the essential design of a protein where this is the result of a cloned quality,
2. The blend of recombinant protein by a reasonable expression system, and
3. The design of DNA tests for use in hybridisation methods. The innovation has driven, e.g., to the development of human insulin for the treatment of diabetics, human growth factors for youngsters with development issues, and an antibody for hepatitis B. DNA tests are fundamental for the location of abnormal genes which are the reason for various illnesses and the strategy has the potential for quality treatment.

S

Saccharase

Any hydrolase that hydrolyzes the fructose (fructosaccharases) or glucose (glucosaccharases) end of proper oligosaccharides, freeing fructose or glucose separately. The term is likewise utilized for invertase (which is both a fructosaccharase and a glucosaccharase).

Saccharin

2,3-dihydro-3-oxobenzisosulfonazole; O-sulfobenzimide; an unnatural, nonnutritive sugar that is a few hundred times better than sucrose. Rodents getting high dosages develop growths of the urinary tract however there is no obvious proof of such an impact in people.

Saccharomyces

A variety of growing yeasts that reproduce asexually by budding or sexually by formation. They are utilized particularly as direct model organisms in the investigation of eukaryotic cell science, and in genetic engineering. *Saccharomyces cerevisiae* is utilized in bread-production and in the development of cocktails and modern liquor. It divides vegetatively in either the haploid or diploid stage. This allows the passive transformations in the haploids and complementation testing in the diploids. Engineering strategies that license DNA take-up by either protoplasts or entire yeast cells have been created and cloning vectors developed. In situations where a recombinant glycoprotein is wanted, yeast has demonstrated extremely helpful since it can, dissimilar to microorganisms, impact posttranslational glycosylation. It has been utilized for the development of an antibody for hepatitis B.

Figure 38. An Illustration of Saccharomyces.

Source: Image by Wikimedia Commons

Saccharopine Dehydrogenase (NAD+, L-Lysine-Forming)

EC 1.5.1.9; a chemical associated with the blend and corruption of lysine. It catalyzes the oxidation by NAD+ of N6 - (L-1,3-dicarboxypropyl)- L-lysine to L-glutamate and 2-aminoadipate 6-semialdehyde with the development of NADH. In bovine and human liver, this movement, with that of lysine-2-oxoglutarate reductase, forms part of the bifunctional aminoadipic semialdehyde synthase. One more saccharopine dehydrogenase utilizes a similar response approach, in which the coenzyme is NADP+: EC 1.5.1.10, saccharopine dehydrogenase (NADP+: L-glutamate-forming). is essential for the bifunctional aminoadipic semialdehyde synthase system.

Safety Cabinet

A cabinet giving a region at seat level in which pathogenic species might be taken care of or put away while forestalling their entrance into the environment outside the area; simultaneously, sterility inside the area is regularly guaranteed. Typically, the cabinet is ventilated with a flood of air, which is separated through a HEPA channel system that traps any microorganisms. Cabinets are ordered by the degree of insurance they offer.

Class I cabinets have an internal progression of air through the front gap giving assurance to staff and the environment.

Class II cabinets are ventilated by filtered air; this is separated to guarantee additionally an internal course through the front gap, the removed air being recycled through the channel and giving security to staff, area, and environment.

Class III cabinets are for all the more possibly risky materials; there is no front gap, the administrator controlling the inside parts utilizing gloves precisely fixed into the front of the cabinet. They give outright all-round security.

Salbutamol

Other name: albuterol; 4-hydroxy-3-hydroxymethyl-a-[tert-butylamino) methyl]-benzyl liquor, a1 - [[1,1-dimethylethyl)- amino] methyl]-4-hydroxy-1,3-benzenedimethanol; a b-adrenergic agonist whose b2 - adrenoceptor movement is multiple times that of isoproterenol.

Salicylic acid

2-hydroxybenzoic acid; a water-solvent derivative of phenylalanine in plants, where it is framed by hydroxylation of benzoic acid. It starts thermogenesis in the voodoo lily (where process it has been called calorigen) and is related with protection from contamination by growths, microorganisms, and infections by stimulating pathogenesis-related proteins. It is the active molecule (as methylsalicylate, saligenin, or their glycosides) accountable for the pain-relieving activity of plants like willow, poplar, and meadowsweet, which have been utilized for this reason over centuries. It is the active element of the extremely adaptable medication acetylsalicylic acid.

Saline

1. Concerned about, or containing normal salt (sodium chloride).
2. Containing, salts of antacid metals or magnesium. 3 a fluid preparation of sodium chloride and (once in a while) different salts, of characterized osmolarity, ready for intravenous infusion or perfusion; typically, 0.9% w/v sodium chloride.

Saliva

The blended emissions of the salivary organs and of the mucous film of the mouth. It contains a-amylase (ptyalin), mucin, different inorganic particles, urea, supports, lysozyme, and so forth, and has a pH close to 7. It is discharged in light of food, which it soaks and greases up; the amylase may likewise hydrolyze ingested starch.

Salivary Gland

Any of the different organs that release saliva into the mouth. In people, there are three sets: the parotid, sublingual, and submaxillary organs. In snakes, they incorporate the poison glands.

Salkowski Test

A test for cholesterol. Whenever concentrated sulfuric acid is added to a chloroform mix of cholesterol, the chloroform layer shows a red to blue tone and the acid layer shows a green fluorescence. [After Ernst Leopold Salkowski (1844-1923), German physiological chemist.]

Salla Disease

An autosomal recessive disease, fairly genetic in Finland, where a variable level of psychomotor disability goes with collection of free sialic acid (and glucuronic acid) in lysosomes and unreasonable discharge of sialic acid in urine. Patients have a close to typical life expectancy. It is brought about by changes in the quality for sialin, and is a milder type of juvenile sialic acid disorders.

Salmon

Any different fish of the family Salmonidae, which live in marine waters of the North Atlantic and North Pacific and breed in waterways. Their tissue is nearly rich in oils with long-chain n-3 unsaturated fats; average composition (significant unsaturated fats, expressed as % of complete unsaturated fats): 16:0, 10; 16:1(n-7), 5; 18:0, 4; 18:1 (n-9), 24; 18:2(n-6), 5; 18:3(n-3), 5; 20:1(n-9), 1; 20:4(n-6), 5; 20:5(n-3), 5; 22:6(n-3), 17.

Salmonella Mutagenesis Test

Or Ames test is a technique for evaluating synthetic compounds for cancer-causing nature by their mutagenic impact on chosen strains of *Salmonella typhimurium* - there being a high relationship among's mutagenicity and cancer-causing nature. The strains typically utilized are those that require histidine. They are plated out on histidine-deficiennt medium subsequent to being exposed with the possible mutagen. Changes instigated by the test substance that re-establish the capacity to incorporate histidine are uncovered by colonies developing on the plate.

Salmon Sperm DNA

It's a preparation of DNA utilized for prehybridization, and hybridization, of films in Northern or Southern blotching of mammalian RNA and DNA. It is picked in light of the fact that its developmental separation from vertebrates makes it improbable to bind DNA tests under the states of hybridization.

Salt Bridge

1. A cylinder, generally shut with permeable fittings, that is loaded up with an answer of a salt (frequently potassium chloride, usually immersed) and used to get electrical contact between two electrolytic half-cells without intermixing of their separate electrolytes.
2. Any electrostatic bond, among decidedly and contrarily charged groups on amino-acid residuesof a protein, that adds to the strength of the protein structure.

Salvage Pathway

Any metabolic pathway that uses for biosynthetic purposes in catabolism. Hence, free purine and pyrimidine bases might be changed over to the relating ribonucleotides. On account of purines, the ribose phosphate moiety of 5-phospho-a-D-ribosyl diphosphate (I) is moved to the purine; hence

adenine + I = adenylate + PPi.

Sandwich Assay

Or sandwich procedure a sort of immunoassay in which the antibody against the antigen to be measured is bound to a strong surface (for example plastic). After treatment of preparation containing antigen the proper antigen is washed. A subsequent counter-acting agent, which is radioactive or fluorescent, is added, sandwiching the antigen; after washing. The subsequent antibody might be specific for an alternate epitope on the antigen, in this manner improving in general particularity, or for the main antibody bound to an antigen.

Sanger Method

1. (for distinguishing and assessing N-terminal amino-acid residues of polypeptides) a strategy wherein free unprotonated amino groups respond with 1-fluoro-2,4-dinitrobenzene (FDNB). The dinitrophenylamino groupes framed are steady to the acid used to hydrolyze the peptide bonds; the yellow arylated (DNP-) amino acids so delivered can then be recognized by chromatography and assessed spectrophotometrically.
2. (for polypeptide sequencing) an overall strategy for determining the essential grouping of a polypeptide chain, in view of specific hydrolytic degradation of the chain into more smaller peptides. The grouping of each more small peptide may then be found by utilization of FDNB to mark its N-terminal amino acid, along with total hydrolysis to give the amino-acid sequence and further stepwise degradation by utilization of carboxypeptidase or by Edman reaction. By choice of degradation techniques to guarantee cross-over of groupings between the different more smaller peptides, the general sequence is settled.
3. (for DNA sequencing) an elective name for the chain-termination technique. [After Frederick Sanger (1918-), British biochemist.]

Sanger's Reagent

2,4-dinitrofluorobenzene (abbr.: DNFB); 1-fluoro-2,4-dinitrobenzene (abbr.: FDNB); a substance utilized in basic protein science to arylate free amino groupings.

Sanitizer

A substance that altogether decreases the bacterial populace in the surrounding however doesn't annihilate or dispose of all microorganisms.

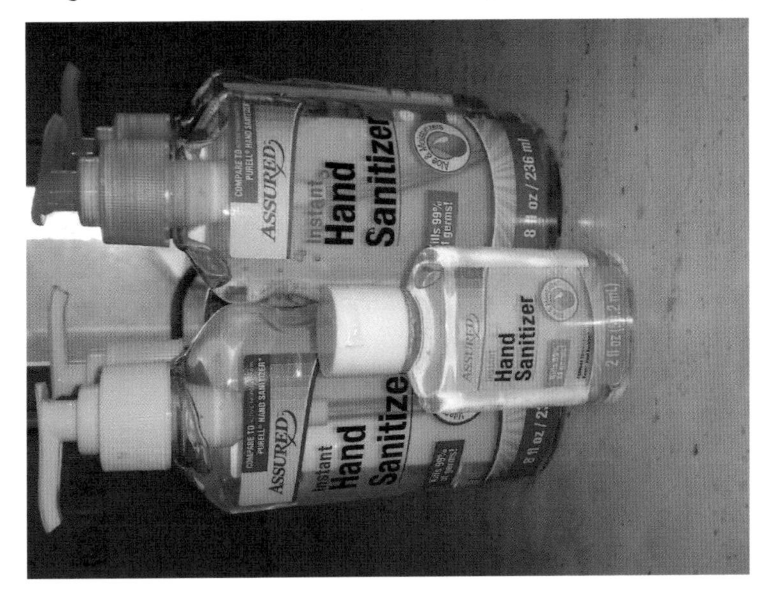

Figure 39. An Illustration of Hand Sanitizer.

Source: Image by Wikimedia Commons

Saponification Number

The number of milligrams of potassium hydroxide consumed in the total saponification of one gram of a specific fat, oil, or wax.

Saponin

Any glycosides, generally found in plants, that are strong surfactants. Every saponin comprises of an aglycon moiety (for example sapogenin), which might be a steroid or a triterpene, and a sugar moiety, which might be glucose, galactose, a pentose or methyl pentose, or an oligosaccharide. All saponins froth emphatically when shaken with water. They are membrane active, strong hemolytic specialists and are utilized at low levels to permeabilize cells. They are exceptionally harmful on injection, but not on ingestion as they are not assimilated through the stomach.

Sarafotoxin

Any poisonous peptides from the tunneling asp, *Atractaspis engaddensis*. They cause heart failure when infused into mice, and are solid vasoconstrictors. All have 21 amino acids, with four cysteine residues in indistinguishable positions and tryptophan at the C terminal. They are homologous with endothelins and have endothelin receptor agonist activity.

Sarafotoxin S6D has the succession CTCKDMTDKECLYFCHQDIIW.

Sarcolemma

The external layer of a muscle fiber. It comprises the plasma layer, a covering cellular membrane (around 100 nm thick and once in a while normal to more than one fiber), and the related free organization of collagen filaments. The term was first used to assign the line apparently denoting the external fiber edge.

Sarcomere

The basic unit of a myofibril of vertebrate muscle, around 2.3 lm long. In striated muscle the sarcomeres of many equal myofibrils are situated to such an extent that myosin thick fibers are adjusted across the myofibril (having a thick appearance in tiny arrangements), as additionally are the actin myofibers (which have a lighter appearance), bringing about the shift of light, isotropic groups known as I groups, and dim, anisotropic groups known as A groups; the I band is divided by an extremely thick, thin Z line, while the focal, less thick area of the A band is known as the H zone, which thusly is separated by the dim M line, or midline, the area of specific proteins that interface contiguous thick fibers to one another. The Z lines are because of connection locales for thin fibers.

Satellite DNA

A small portion of DNA, adding up to 10% or a greater amount of total DNA in most eukaryotic cells, with a consensus nucleotide succession. Its base synthesis leads to a 'satellite' band having an alternate light thickness when DNA preparations, following shear force application, are centrifuged in cesium chloride gradients. Mouse satellite DNA contains around 1,000,000 duplicates of a repeat of 300 bp with a lower (G+C) content than that of the regular DNA arrangements.

Saturation

1. (of a substance compound) where all the valence rules of the atoms are fulfilled, particularly a natural compound that contains just single carbon-carbon bonds.
2. (of a solution) the state where it has the best concentration of the solute that can stay in stable equilibrium with undissolved solute at a given temperature and pressure.
3. (of a gas) the state where there is the best convergence of the gas related with a fluid or strong that can stay in stable equilibrium with unvaporised fluid or strong, individually, at a given temperature and pressure.
4. (of a functioning site of a chemical or carrier) the place where the site is completely filled with the ligand. Saturable chemical or receptor systems show hyperboliform or sigmoid bends of rate versus substrate.
5. The demonstration or cycle of bringing to or towards saturation; particularly the increase of hydrogen to significantly increase and twofold bonds.

Scanner

An instrument or gadget used to quantify or test the appropriation of some amount or condition in a specific system, region, or area, for example, to quantify color or radioactivity on a chromatogram or electrophoretogram, or to gauge the outspread dissemination of light absorbance in a cell in an ultracentrifuge.

SCAP

Abbreviation for SREBP-Is a cleavage initiating protein; a layer protein of the endoplasmic reticulum and the Golgi system that manages cleavage and arrival of the transcriptional area of SREBP by S1P and S2P acting in succession. It comprises of N-terminal transmembrane fragments (of which five establish a sterol-detecting area), and a cytoplasmic C-terminal end that contains five WD40 repeats and interfaces with the functional space of SREBP. Cholesterol and polyunsaturated unsaturated fats, acting by means of the sterol-detecting area, repress cleavage of SREBP and arrival of its transcriptional space.

Schiff Base

Or Schiff's base any imine having the more restricted general construction R2C=NR′ where R might be any organyl group or H and R′ is any organyl group. It could be formed by build-up of the carbonyl group of an aldehyde or ketone with the amino group of an essential amine as per the condition:

R2C=O + H2NR′ = R2C=NR′ + H2O

Regularly viewed as inseparable from azomethine.

Schizosaccharomyces Pombe

The yeast that is utilized to blend the African brew called pombe. The pole formed cells develop by prolongation, and the spores bud and separate as haploids. The membranous organelles, for example, the Golgi mechanical assembly, are intently like those of higher eukaryotes, in vast difference to those of maturing yeast. It is reasonable for investigating the connection between cell size and the cell cycle since: (1) the cells partition evenly, in contrast to maturing yeast; and (2) the yeast grows long however not in measurement.

Schlesinger Test

A subjective test for urobilin in urine. The urobilinogen in the urine is first oxidized by iodine to urobilin, then, at that point, a zinc-urobilin complex is formed by reaction of zinc acetic acid derivation in ethanol; the complex has a yellow-green fluorescence and an absorption band at 507 nm.

Schlieren Method

An optical procedure for showing inhomogeneities in the refractive list, reliant upon the bending of a ray of light from its undisturbed way when it goes through a medium where with changes in density. The shape of the ray is corresponding to the refractive angle toward the path typical to the ray. It is utilized in organic chemistry especially in estimations of dispersion, sedimentation, and electrophoretic versatility of proteins and different substances.

Schmidt-Thannhauser Procedure

A method for the extraction and determination of DNA, RNA, and phosphoproteins. Material insoluble in dilute, cold trichloroacetic acid is dried, processed with salt, and along these lines precipitated with a mineral acid. RNA is solubilized to acid-solube nucleotides while DNA resists the process to be precipitated with mineral acid.

Schultz-Dale Reaction

An *in vitro* anaphylactic response wherein the uterus or ileum of a sensitized organism (guinea pig) contracts specificaly when a modest quantity of the sensitizing antigen is added to the washing medium. The response is brought about by receptor and comparable substances to scan for hypersensitivity.

Schütz-Borrisow Rule

Or Schütz regulation is an experimental rule as per which the speed of a compound response is relative to the square foundation of the protein center. It was first produced for pepsin however applies just to the unrefined catalyst under specific restricted conditions; comparable outcomes have been gotten for certain different proteinases. The impact is because of the presence of a reversibly separating pepsin inhibitor in unrefined pepsin preparations.

Scintillation Cocktail

The radioactivity of a sample is measured by adding a liquid scintillator that releases alpha and beta particles for detection. The first scintillator to be generally utilised was 2,5-diphenyloxazole (PPO), frequently with 1,4-bis(5-phenyl-2-oxazolyl) benzene (PoPoP), which is invigorated by the light produced by PPO to transmit light of a higher frequency more appropriate for estimation. Numerous other scintillators, particularly biodegradable mixtures and others more viable with water, have been created.

Scintillation Counter

A scintillation counter is an instrument for detecting and quantifying ionizing radiation by using the excitation effect of incident radiation on a scintillating material and detecting the resultant light pulses. It has a scintillator to generate photons in response to incident radiation, a sensitive photodetector (commonly a photomultiplier tube or PMT), a charge-coupled device (CCD) camera, or a photodiode) to convert light to electricity and subsequent processors for signal analyses.

Scoring Matrix

Or weight grid (in preparation investigation) a table of all conceivable pairwise connections between nucleotide or amino acid images, utilized by database hunt and grouping preparation projects to measure the likeness between successions. The easiest scoring network is a unitary system, where just pairwise characters get a positive score (nonidentical characters get a zero score).

Scotophobin

A 15-residue peptide that collects in the brains of rodents prepared to keep away from the dark; the organisation is said to get a comparative reaction in undeveloped organisms. It is believed to be the main memory-coordinating substance.

Selenium

A metalloid component of group 16 of the (IUPAC) intermittent table; nuclear number 34; relative nuclear mass 78.96. Its primary oxidation states are - 2, +2, +4, and +6, with properties like those of sulfur, yet at the same more metallic. It occurs in different allotropic structures (dim, red, dark); the dark structure shows some electrical conductivity, which is upgraded by light, and it is utilized in 'selenium cells' to quantify light force. The most bountiful isotope is selenium-80 (mass 79.916; relative overflow 49.8%); others are selenium-74 (0.9%), selenium-76 (9.0%), selenium-77 (7.6%), selenium-78 (23.5%), and selenium-82 (9.4%). Radioactive isotopes are selenium-72, selenium-73, selenium-75, and selenium-79. Selenium is a fundamental minor element, expected for the development of selenoproteins, remarkably glutathione peroxidase. Stomach absorption is poor, yet selenium deficiency is interesting (seen, e.g., in certain parts of China with low soil selenium), and appears as myopathy (particularly cardiomyopathy). Estimation of red-cell glutathione peroxidase gives a record of selenium status.

Figure 40. An Illustration of Selenium in Sandstone

Source: Image by Wikimedia Commons

Semaphorin

(Abbreviation as Sema) or collapsin any of a group of emitted or membrane bound proteins (≈750 deposits) that are expressed in different embryonal and adult tissues. In the developing sensory system, they are delivered by cells that encompass the ways of axon relocation and can repulse (for example Sema 3A and Sema 3F) or attract developing axons. They may likewise have non-neuronal capacities. Their extracellular N-terminal areas (≈500 residues) are homologous to those of hepatocyte development factor receptors and of plexins. Semaphorins are overexpressed during metastatic movement of cancers. Plexins are parts of semaphorin receptor structures.

Sequence Analysis Package

A set-up of projects for the computational investigation of protein and additionally nucleic acid preparations, including, for instance, devices for piece group, planning, interpretation, programmed and manual succession arrangement, phylogenetic examination, data set looking, design acknowledgment, property perception, organization examination, and RNA plotting. One of the main packages was GCG, which then, at that point, turned into a commercial item; a public option is EMBOSS, the European Molecular Biology Open Software Suite.

Sorbitol

A non-systematic name for D-glucitol, one of the ten stereoisomeric hexitols. It can be obtained from glucose artificially or metabolically (by NADPH-requiring aldehyde reductase; other name: aldose reductase, EC 1.1.1.21) by decrease of the aldehyde group, and is changed over to D-fructose by L-iditol 2-dehydrogenase (EC 1.1.1.14). It is broadly present in green algae and higher plants, and was first found (1872) in the juice of berries from mountain debris (*Sorbus aucuparia*). It is about half as sweet as sucrose, and is frequently utilized in sugars related to saccharin. Economically it is critical in the food, drug, paper, and material enterprises. Its preparation from excess glucose centers in diabetics might have neurotic ramifications in tissue, which can't change it over to fructose.

Subunit-Exchange Chromatography

A type of chromatography where protein subunits are immobilized on a strong system and permitted to react with protein subunits in preparations. Evaluation of subunit trade between the network and preparation might be utilized in examining the affiliation separation properties of the protein system. The strategy additionally gives a strong, specific separation technique.

Succinate Dehydrogenase

EC 1.3.99.1; precise name is succinate (acceptor) oxidoreductase; different names are fumarate reductase/dehydrogenase. A bacterial compound, or a corrupted substance from succinate dehydrogenase (ubiquinone), a significant part of mitochondrial complex II. It catalyzes the oxidation of succinate by an acceptor to shape fumarate and a reduced acceptor; it doesn't respond with ubiquinone or free FAD.

Swaposin

An element of certain proteins where domains, naturally found in one protein, are tracked down organized in an alternate order in different proteins. The peculiarity was first remarked on comparable to specific plant aspartic peptidases, which contain saposin-like spaces exchanged into an alternate order from that found in saposins (consequently the name). In another model, some pleckstrin homology domains are viewed as held by additions of SH areas.

T

Tachykinin

Any of a group of closely related vasoactive peptides portrayed by a fast impact on vascular (and extravascular) smooth muscle, creating hypotension. They likewise directly affect sensory tissue, and activate salivary and lachrymal emission. Natural tachykinins are oligopeptide amides containing 10-12 amino-acid residues, and ending in the normal sequence - Phe-Xaa-Gly-Leu-Met-NH2 (where Xaa = Phe, Tyr, Ile, or Val). Mammalian tachykinins include neuromedin K, neurokinin A (previously substance K) and substance P; others (e.g., eledoisin, physalaemin) occur in organisms of land and water and octopods. Substance P and neurokinin An are produced from a preprotachykinin protein precursor. The essential mRNA record is differentially handled into three mature mRNAs: a, b, and c; the b and c preprotachykinin mRNAs are prevalent in the rodent, each encoding substance P and neurokinin A.

Tandem Affinity Purification

Abbreviated as TAP which Is a detachment strategy intended to seclude protein edifices, by which an IgG-restricting space grouping tag is added to the quality for an objective protein, and the protein is blended with the segment encoded by the tag. Proteins with the space are isolated from a blend utilizing a partiality section containing IgG globules, to which they stick. Assuming the objective protein is important for an intricate, proteins related with it are additionally held on the section.

Tangier Disease

An uncommon autosomal recessive condition in people wherein there is an inadequacy or complete shortfall of high-density lipoprotein. Homozygotes have exceptionally low degrees of apolipoprotein A-I. There is group of cholesterol esters in cells of the mononuclear phagocyte system. The condition is generally harmless as organ function isn't impacted.

Taurocholate

The anionic type of taurocholic acid, N-cholyltaurine, a significant bile salt of people and different vertebrates. It is valuable as an anionic bile-salt emulsifier (total number 4; CMC 3-11 mM). Taurine forms of other bile acids are known, for example, taurochenodeoxycholate and taurodeoxycholate.

Taxol

An antitumor and antileukemic molecule isolated from the bark of the Pacific yew, *Taxus brevifolia*. It is utilized clinically to treat unmanageable ovarian malignant growth. It hinders mitosis and stalls the cell cycle at G or M stages through an upgrade of the polymerization of tubulin and the ensuing adjustment of the microtubules. Taxol is made semisynthetically from a compound present in *T. baccata*. An absolute synthetic reaction has additionally been accomplished. A compound with comparable properties, separated from the myxobacterium *Sporangium cellulosum*, is epothilone; this is more water-soluble than taxol and can be created in amount by maturation. Taxol and its analogs are known as taxanes. One exclusive name is Paclitaxel. An exclusive name of a semisynthetic simple is Docetaxel.

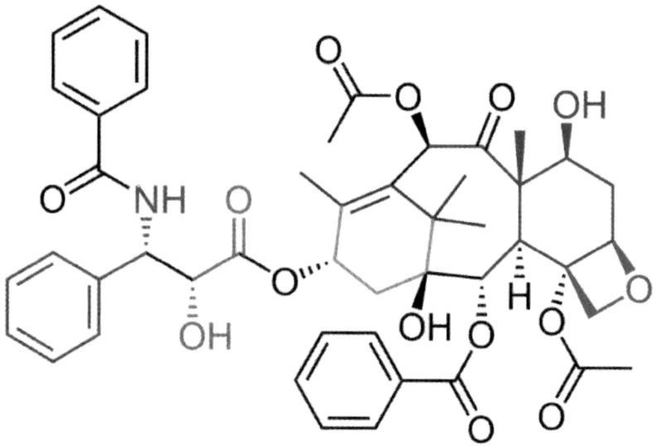

Figure 41. An Illustration of Taxol Total Synthesis.

Source: Image by Wikimedia Commons

Tay-Sachs Disease

Or GM2 gangliosidosis an acquired neurological metabolic issue because of lack of the lysosomal protein, hexosaminidase (structure A, lack in a chain). This prompts the accumulation of its substrate, ganglioside GM2, in the brain. The disease is generally normal (1 out of 2500 live births) among Ashkenazy Jews. There is motor and mental damage with an augmented reaction to sound; by year and a half old enough, most patients are visually impaired, hard of hearing, and spastic; they die by the age of 3 years.

TC System

The classification of layer transport proteins, involving a five-digit assignment for every one of the characterized and sequenced transport proteins. It includes a few classes: 1. Pores and channels; 2. Electrochemical-potential-driven transporters; 3. Primary active transporters; 4. Group translocators; 5. Transmembrane electron transporters; 6. Accessory factors engaged with transport; 7. Incompletely characterized transport systems), each class containing a few subclasses; the subclasses are partitioned into families, the families into subfamilies and these then contain every one of the applicable individual proteins. Along these lines, for instance, bacterial Ag+-exchange ATPase is TC 3.A.3.5.4.

T-DNA

Abbreviation for move DNA; that section (23 kb) of the Agrobacterium Ti plasmid that is consolidated into the genome of infected plant cells. It codes for a nopaline or octopine, which the altered cells produce and the organisms then, at that point, use. Each end of the T-DNA is flanked by an almost indistinguishable 25 bp repeatt; the righthand one is fundamental for the exchange cycle.

Teichoic Acid

Any polymer present in the cell wall, film, or vacuole of Gram-positive microorganisms and containing chains of glycerol phosphate or ribitol phosphate deposits. Moreover, starches are connected to glycerol or ribitol, and some - OH groups are esterified with D-alanine.

Tektins

A group of proteins that are in eukaryotic cell cytossskeletal structures. Tektins have comparative amino-acid sequences: they include actin, actin-like proteins, some microtubule proteins, mitochondrial core protein, and erythrocyte membrane proteins.

Telomerase

Also known as telomerase invert transcriptase (abbreviation as TERT) is a DNA polymerase protein associated with telomere development, and in telomere lengthening to keep up with the telomere arrangements during replication. The subunit contains a transcriptase and one telomerase-specific theme. Action is imperceptible in most human physical cells yet is available in microbial cells, in ≈90% of growths, and in altered cell lines. The molecule locates the G-rich strand of a current telomere sequence, extending it in the 5′ to 3′ course. The yeast *Candida albicans* contains two profoundly homologous subunits every one of 867 amino acids. In parting yeast, the centre chemical contains Est2p before the S stage, and becomes initiated late in the S stage when Est1p and Est3p are much expressed. The RNA part facilitates the G-rich repeat grouping, and in humans is limited by dyskerin.

Telomere

The DNA-protein structure that seals either end of a chromosome. Telomeric DNA comprises of basic tandemly repeated sequences specific for every species. In human germline cells, the 3′ end of every chromosome contains 1000-1700 repeats of the hexanucleotide TTAGGG, and the 5′ end contains the correlative rehashes. In many cells, the recurrent number lowers with each round of replication. Telomeric repeats are added to the 3′ end of chromosomal DNAs by telomerase, along these lines keeping up with telomere length.

Telopeptide

Any peptide that overhangs from the triple-helical group of tropocollagen, and have a design unique in relation to the triple-helical bits. They are eliminated by proteases, with corresponding breakage of intramolecular interchain bonds. They involve close terminal positions. The antigenic reaction to infused heterologous tropocollagen is guided against these telopeptide arms outside to the triple helix.

Temperate Phage

A bacteriophage that, following disease of its host bacterium, may either enter the lytic cycle, similar to a harmful phage, or lay out an advantageous connection with the host cell that outcomes in its propagation in the relatives of the bacterium. Microscopic organisms transporting a temperate phage are supposed to be lysogenic, while the transported phage is known as a prophage.

Temperature Coefficient Of A Reaction

Image: Q10; the element by which the speed of a substance response is expanded by raising the temperature, t, by 10°C. Normally it is the proportion of the rate of a reaction at (t + 10°) to that at t, signified by Q10 = kt+10/kt , where kt+10 and kt are the rate constants at the separate temperatures. The Q10 value of homogeneous substance responses is for the most part in the reach 2-3; numerous chemical catalyzed responses and physiological cycles display, over a restricted scope of temperatures, a Q10 value of ≈2.

Temperature-Sensitive Mutation

A mutation that is manifest in just a restricted scope of temperature. The variant transporting such a mutation acts typically at the lower, lenient temperature, however, when developed at the lower or higher, prohibitive temperature, it shows the mutant form.

Template

A macromolecule whose construction fills in as an example for the blend of another macromolecule. The term applies especially to a nucleic acid on which another nucleic acid of reciprocal base arrangement is orchestrated, as in hereditary record, or on account of messenger RNA, on which a polypeptide is made (genetic code).

Terminal Deoxynucleotidyltransferase

Abbreviated as TdT; EC 2.7.7.31; suggested name: DNA nucleotidylexotransferase; methodical name: nucleoside-triphosphate:DNA deoxynucleotidylexotransferase; different names are : terminal transferase; terminal deoxyribonucleotidyltransferase. A chemical, usually separated from calf thymus, that catalyzes the progressive exchange of deoxynucleotide residues from the comparing nucleoside triphosphates to the 3′ finish of an oligo-or polydeoxynucleotide. It is a template-free, Mg2+-dependant DNA polymerase.

Tertiary

1. (in science) a prefix: tert- (abbreviation as t-); depicts an alkyl compound (for example an alkanol) in which the useful set (for example a hydroxyl group) is appended to a carbon atom connected to three others. b depicting the carbon atom bearing the functional group in such a compound.
2. Portraying an amide or an amine in which three groups are appended to a nitrogen molecule.
3. Depicting a salt formed by supplanting every one of the three of the ionizable hydrogen molecules of a tribasic acid by one, two, or three different cations of suitable valency.

Tertiary Structure

The degree of protein structure at which a whole polypeptide chain has collapsed into a three-dimensional structure. The tertiary construction results from bonds between amino-acid residues that might be broadly isolated in the whole structure, yet might be brought into vicinity by the folding of the polypeptide chain. The powers of communication between residues incorporate hydrogen bonds and electrostatic attraction. The structure is settled in certain proteins by disulfide bonds. In multichain proteins, the term tertiary design applies to individual chains.

Testicular Feminization

Or called as testicular feminisation is an acquired condition in which hereditary males have formative irregularities going from a total outside female aggregate to some degree masculinized questionable genitalia. The commonest form is X-linked and results from any of a few hundred transformations in the androgen receptor that lower its binding to the chemical. An autosomal recessive form results from numerous transformations that produce steroid 5a-reductase 2 inadequacy and the absence of 5a-dihydrotestosterone.

Tetanus Toxin

The protein toxin of *Clostridium tetani*, the causative agent of tetanus. It is a heterodimer of light and heavy chains. It blocks brain exocytosis and subsequently norepinephrine release, perhaps because of a Zn2+-dependant proteolysis of synaptobrevin by the light chain, yet additionally by the hindrance of a neuronal transglutaminase (protein-glutamine c-glutamyltransferase, EC 2.3.2.13). The protein is integrated as a precursor of 1314 amino acids. This is cut to yield the light chain (deposits 1-456) and heavy chain (residues 457-1314).

Tet-Off

An exclusive name for a system for directing the expression of transgenes *in vivo* by controlling the level of the antibiotic. The quality of interest is set downstream from tetO antibiotic medication reaction components. The antibiotic medication trans-activator tTA (a combination protein joining the N-terminal 1-203 amino acids of the antibiotic medication repressor protein TetR and the C-terminal 127 amino acids of the HSV VP16 activation domain) binds to tetO reaction components. Increase of antibiotic medication or doxycycline causes separation of tTA from the operator along these lines switching off the gene of interest.

Tetra+

Or (before a vowel) tetr+ comb. structure

1. Meaning four, fourfold, every fourth; see likewise tetrakis+.
2. (in synthetic classification) (recognize from tetrakis+) a demonstrating the presence in an atom of four indistinguishable indicated unsubstituted groups; for example, tetrachloroethylene, tetrameth-

ylammonium chloride. b showing the presence in a particle of four indistinguishable inorganic oxoacid residues in straight anhydride linkage; for example, sodium tetraborate.

Tetracycline Resistant

Abbreviation as TCr; is impervious with the impacts of the anti-infection antibiotic medication. Three unique systems of antibiotic medication resistance are known - antibiotic efflux, ribosome targeting, and antibiotic adjustment - and more than 60 antibiotic opposition genes have been sequenced. In intestinal microorganisms, two transposons, Tn1721 and Tn10, regularly transport the genes for antibiotic medication resistance. Tn10 encodes two proteins: a film found protein (TetA, Mr 36 000) that seems to intervene resistance and a repressor protein, TetR (see antibiotic medication repressor), related with the inducibility of antibiotic resistance. The film protein siphons the antibiotic out of the cells against a concentration gradient. New DNA might be joined to the antibiotic resistance gene in a cloning vector consequently empowering bacterial clones containing the DNA to be chosen from those lacking protection from the anti-microbial. The plasmid pBR322 presents protection from both tetracyclin and ampicillin. The addition of new DNA at specific destinations inactivates either or the two genes.

Thermal Denaturation Profile

A curve relating some actual property of a biopolymer in arrangement (for example thickness, optical absorbance, or optical turn) to temperature. The state of the curve demonstrates the event of actual changes in the biopolymer, for example, partition of the two strands of nucleic acid or modification in the tertiary design of a protein.

Thermodynamics

The investigation of the regulations that control the transformation of energy starting with one structure then onto the next, the heading in which heat will move starting with one system then onto the next, and the accessibility of energy to function. The zeroth law of thermodynamics expresses that in the event that two systems are each in thermal balance with a third system, they should be in thermal balance with one another. The principal law of thermodynamics or the law of preservation of energy determines that in an isolated, adiabatic system the amount of the internal energies, all things considered, U, stays consistent, anything that might be simply the progressions inside the system. Assuming an arrangement of consistent mass increases heat, q, and has work, w, done on it then an adjustment of its interior energy, DU, happens with the end goal that $\Delta U = q + w$. (Where the system loses heat, $q < 0$, and where work is finished by it, $w < 0$.) The second law of thermodynamics is about the limitations toward energy stream in normal systems and assesses the irreversibility of regular cycles. It determines that the entropy of any system, S, in contrast to energy, isn't saved yet increments with time in any unconstrained cycle in an isolated system. In the instance of a reversible interaction, the increment in entropy of a system, dS, is characterized as the heat consumed by the system, dqrev, isolated by the thermodynamic temperature, T; for example, $dS = dq_{rev}/T$. The third law of thermodynamics is about the way that the distinction between the change of enthalpy, ΔH, and the change of the free energy, ΔG, of a system decreases as the temperature of the system lessens, and the worth of $(\Delta H - \Delta G)$ will in general zero as the thermodynamic temperature will in general zero. This expects that the adjustment of the heat limit of the system (at consistent strain), ΔC_p , and ΔS should fall to zero at 0 K.

Thin-Film Dialysis

A sort of dialysis that keeps up with the diffusate and the retentate in thin layers in contact with the dialysis film to decrease the general rate of the reaction made by dissemination of solutes in free arrangement. Extremely quick movement or exchange of little atoms can happen, particularly if the volume proportion of diffusate to retentate is high. With fitting membranes, the method may likewise be utilized for insightful investigations of the size and adaptation of peptides and proteins in structures.

Thin-Layer Chromatography

Abbreviated as TLC is a chromatography on a thin layer of any of different solid materials, including silicic acid, aluminum oxide, or cellulose. The layer, naturally somewhere in the range of 0.05 and 1 mm thick, is made to stick to a sheet of glass, aluminum, or plastic. Thicker layers acknowledge heavier stacking, while more thin layers settle all the more productively and quickly. Settling power is high, runs are fast, and, in light of the fact that help materials, for example, silicic acid or aluminum oxide are dormant to them, acid treatments, like sulfuric acid, can be utilized to change natural materials over to carbon (an interaction known as scorching), uncovering them as dark spots.

U

UBE3A

This is the symbol for the gene that encodes the ubiquitin protein ligase E3A, which is responsible for binding the tiny protein ubiquitin to proteins that are targeted for breakdown in proteasomes. In some cases of Angelman syndrome, mutations in the gene are responsible for the condition. The paternal allele has been imprinted on the child.

Ubiquilin

An enzyme that interacts with presenilins and increases their accumulation while also appearing to boost their synthesis. Its structure is similar to that of ubiquitin, and it, like ubiquitin, is found in the neuropathological lesions that are diagnostic of Alzheimer's disease (neurofibrillary tangles) and Parkinson's disease (amyloid plaques) (Lewy bodies). The relevant gene is located in a region that is thought to include a gene for late-onset Alzheimer's disease, according to researchers.

Ubiquinol–Cytochrome C Reductase

EC 1.10.2.2; systematic name: ubiquinol:ferricytochrome-c oxidoreductase; other name: complex III; formal name: ubiquinol:ferricytochrome-c oxidoreductase. In the presence of two molecules of ferricytochrome c, an enzyme complex that catalyzes the oxidation of ubiquinol to generate ubiquinone and two molecules of ferrocytochrome c is formed. It contains the cytochromes b562, b566, and c1 as well as a Rieske iron–sulfur protein, among other things.

Ubiquinone or (formerly) Coenzyme Q

In short, Q is a lipid-soluble electron transporting coenzyme that is a key component of the respiratory chain. It is generated in humans from tyrosine, and it is also synthesized in bacteria and other species through the shikimate pathway, which is found in bacteria.

Ubiquitin 1

An amino-acid protein of 76 amino-acid residues that was originally isolated from bovine thymus but has since been discovered in the cells of all tissues investigated, including mammals, yeasts and bacteria as well as higher plants. In both nuclei and cytoplasm, it receives its name from the extensive distribution of which it is a part. In a protein-dependent reaction, it results in the condensation of its terminal with lysine amino groups; this event is mediated by a massive multiprotein complex known as the 26S proteasome, which is composed of over 100 proteins. For example, ubiquitination is responsible for the destruction of cyclin, which is essential in the management of the cell cycle. The cell, therefore, has two mechanisms for protein degradation: ubiquitination in the nucleus and proteolysis in the lysosomes. Modified proteins breakdown in a short period of time. In the laboratory, ubiquitin is generated as a polyubiquitin precursor, which contains precise head-to-tail repeats, with the amount of repeats varying from species to species. In some species, there is the last amino acid after the last repetition of the sequence (e.g., Val in humans). Some ubiquitin genes encode a single copy of ubiquitin linked to a ribosomal protein, while others encode several copies of ubiquitin.

UDP glucose 4-Epimerase

Other names for the isomerase EC 5.1.3.2 include UDPgalactose 4-epimerase (abbreviated as GALE), galactowaldenase (previously known as galactowaldenase), and uridine diphosphoglucose 4- epimerase (abbreviated as UDPgalactose 4- epimerase). In this mechanism, a NAD+ molecule is tightly bound to the hexose moiety of either uridine diphosphosugar or uridine diphosphosugar is oxidized at the C-4 position to form an oxo group, and the resulting NADH is then reoxidized by the intermediate oxo derivative, with the configuration of the hydrogen and hydroxyl groups at C-4 inverted. In the case of mutations that cause enzyme deficiency, the consequence is either a benign abnormality affecting just blood cells or a severe but rare type of galactosemia affecting all organs. The degree of molecular instability caused by the mutation may be reflected in the variation in results. The human gene encoding 348 amino acids has 348 amino acids.

Ultrasound

With frequencies more than 20 kHz, propagated waves of the same type as sound but inaudible, are produced. In addition to being employed for dissolving cells and subcellular components, frequencies of 25 kHz can also be used in cleaning baths to remove residues from a wide variety of materials. Heat is produced in the range of 0.5–3 MHz, and this heat is used to relieve sore muscles by increasing the permeability of cell membranes, which allows fluid to be removed from the muscle tissue. The frequency range of 3–12 MHz is often utilized for diagnostic purposes, and it is a popular way of non-invasive imaging, particularly in pregnant women. There is also ultrasonic microscopy, which is similar to light microscopy but uses sound in the GHz range to achieve its results. Ultrasound is also being researched and developed for the purpose of disintegrating tumors.

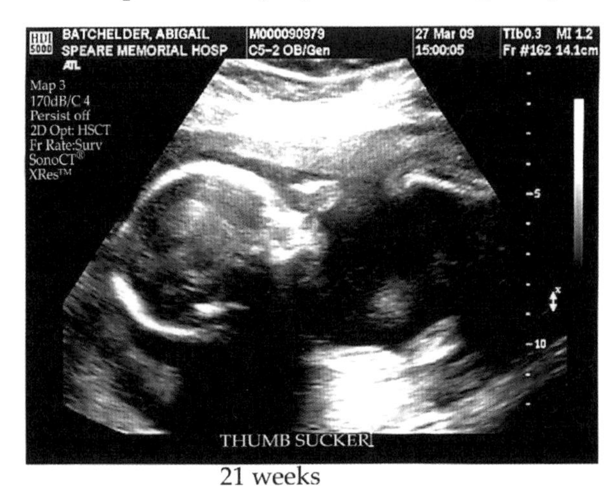

Figure 42. An Illustration of Ultrasound.

Source: Image by Flickr

Ultraviolet (radiation)

UV or UV is an abbreviation for ultraviolet radiation, which is electromagnetic radiation with wavelengths ranging from 13.6 to 400 nm (frequency range: 750 THz to 22.1 PHz). In the visible spectrum, this range includes wavelengths that overlap with those of soft X radiation and ends just beyond the short-wavelength limit of visible light. UV radiation is commonly split into two categories: near-ultraviolet (200–400 nm) and far-ultraviolet (or vacuum ultraviolet). Because far-ultraviolet radiation is absorbed by oxygen, it is necessary to utilize evacuated apparatus or to flush with N2 when working in the far ultraviolet.

Unconventional Myosin

Any of the approximately 14 isoforms of muscle myosin (myosin II) that operate as motor proteins and are found in a variety of cell types other than muscle. Each molecule is composed of a globular head (which contains an actin-activated ATPase) and a short tail (which binds membranes including those of organelles and vesicles). Cytokinesis, cytoplasmic streaming, and axonal transport are all facilitated by the movement of the head over actin microfilaments. Myosin I is found in the majority of nonmuscle cells. When melanosomes are transported in pigment cells and vesicles are transported in neurons, myosin Va is responsible. Griscelli syndrome is caused by mutations in the GRSC gene located on chromosome 15q21 (hypopigmentation of skin and hair and immunological deficiency). Molecular weight: 2215 amino acids. Myosin VIIa is an amino acid protein. Usher syndrome type I is caused by mutations in the gene at 11q13.5, which is connected with a kind of familial nonsyndromic deafness, whereas others are caused by mutations in the gene at 11q13.5 (congenital deafness with vestibular dysfunction and retinitis pigmentosa). Myosin XV is a protein with a length of 3530 amino acids that is found in the sensory cells of the inner ear. Mutations in its gene, which is located on chromosome 17p11.2, cause a kind of familial nonsyndromic deafness.

Uncoupling

It is the dissociation of phosphorylation from oxidation in the respiratory chain that allows electron transport to proceed without the necessity for esterification of phosphate or the use of inorganic phosphate (in biochemistry).

An uncoupling agent (also known as an uncoupler) is a substance that causes the uncoupling of phosphorylation from electron transport at one or more places in the electron-transport chain. Uncoupling agents (also known as uncouplers) are used to treat a variety of conditions. Dinitrophenol is a classic example of this. Inhibitors of oxidative phosphorylation most likely work by causing protons to leak through the mitochondrial inner membrane, disrupting the proton gradient that is necessary for oxidative phosphorylation to occur.

Undulin

A non-collagenous glycoprotein that is found in the interstitial extracellular matrix and that is linked to collagen as bundles of homogeneous wavy fibers in mature collagen fibrils, it is found mostly within tightly packed mature collagen fibrils when isolated from skin or placenta. In terms of its relationship to collagen type XIV, it is a member of the fibronectin–tenascin family of proteins. Soft tissue dense collagen matrices include this protein, which is thought to have a role in the supramolecular organization of interstitial collagens. Undulin 2 in humans has a sequence that is comparable to that of fibronectin type III.

Uniport

Membrane transport is a process in which a single type of substrate (ion or uncharged molecule) is transported across the membrane, either by the mechanism of mediated diffusion, where the substrate equilibrates at the two membrane sides (e.g., monosaccharides in mammalian erythrocytes; TC 2.A.1.1.12) or by the mechanism of active transport, where an external source of energy is used to achieve movement against a chemical potential or an electrochemical Uniporters are transportation systems that only move one molecule.

Unit Membrane

The term "membrane" can refer to any membrane, regardless of its cellular location, that when stained with osmium or other heavy metals appears in section in the electron microscope as a pair of parallel dark lines separated by a less dense layer, the entire structure being 6–10 nm thick. The plasma membrane, as well as the membranes of organelles such as mitochondria, nuclei, and the endoplasmic reticulum, are examples of unit membranes.

Universal Primers

DNA sequencing primers that are specific for the plasmid sequences flanking the cloned DNA insert are oligonucleotides that are employed as primers in DNA sequencing processes. This position is frequently occupied by promoters such as the T3, T7, or SP6 promoters, which can be targeted with primers that anneal to them in order to obtain sequenced inserts that have been cloned into a variety of vectors.

Uric Acid

In certain mammals (including humans), 2,6,8-trioxypurine (also known as purine-2,6,8-triol) is the end product of purine metabolism, and in uricotelic animals, it is the primary nitrogenous excretory product. It is produced by the enzyme xanthine oxidase from xanthine. Gout is caused by the accumulation of monosodium urate crystals in the tissues, which is due to the fact that uric acid is only marginally soluble in water.

Ursodeoxycholate

In its conjugate form with taurine, ursodiol 3a,7b-dihydroxy-5b-cholan-24-oate (or ursodiol 3a,7b-dihydroxy-5b-cholan-24-oate) is discovered in the bile of bears (thus the name, which comes from the Ursidae family). It works by preventing the synthesis and absorption of cholesterol, and it has been shown to aid in the dissolving of gallstones in some cases.

Uroporphyrin-III C-Methyltransferase

Uroporphyrin-III C-methyltransferase (abbreviated SUMT) is an enzyme with the EC number 2.1.1.107 and the systematic name S-adenosyl-L-methionine:uroporphyrin-III C-methyltransferase (abbreviated SUMT). Two molecules of S-adenosyl-L-methionine combine with uroporphyrin III to generate two molecules of S-adenosyl-L-homocysteine and one molecule of sirohydrochlorin when the enzyme catalyzes the reaction. It is a component of the enzyme siroheme synthase, which is involved in the manufacture of the antioxidants siroheme and cobalamin.

Usher Syndrome

Depending on the type of syndrome, Usher syndrome can be associated with a variety of hearing disorders, including vestibular dysfunction and congenital deafness, partial hearing loss, and progressive hearing loss. Type I of the syndrome is associated with vestibular dysfunction and congenital deafness, type II with partial hearing loss, and type III with progressive hearing loss. A total of six genes are associated with subtypes of type I (myosin VIIA in type IB and harmonin in type IC); one gene is responsible for type II and encodes an extracellular matrix protein (approximately 1500 amino acids) found in the retina and other tissues; one gene is responsible for type III; and one gene is responsible for type IV.

U-snRNA

U-snRNA is an abbreviation for uridylate-rich small nuclear RNA; it is a family of small nuclear RNA molecules that include 100–215 nucleotides and has a 'capped' 5′ end. U-snRNA molecules are found in the nucleus of bacteria and in the nucleus of plants. They are most commonly found as tiny nuclear ribonucleoprotein particles in the nucleus. It is possible that their role is connected to the splicing of heterogeneous nuclear RNA (hnRNA) into mature messenger RNA (mRNA).

Uteroglobin or Blastokinin

A steroid-inducible secreted protein known as uteroglobin or blastokinin, which regulates progesterone concentrations reaching the blastocyst and is a powerful inhibitor of the enzyme phospholipase A2. It is a homodimer that is disulfide-linked.

UTP–hexose-1-phosphate uridylyltransferase

UTP–hexose-1-phosphate uridylyltransferase is a protein with the systematic name UTP–hexose-1-phosphate uridylyltransferase. Uridinediphosphogalactose (UTP) is a sugar that can be converted into uridine diphosphogalactose by the action of an enzyme called UTP. It is responsible for catalyzing the reaction between UTP and a-D-galactose 1-phosphate, which results in the formation of UDPgalactose and pyrophosphate. Sugar substrate 1-phosphate can likewise act as an acceptor in the formation of uridine diphosphoglucose (albeit at a slower pace than glucose 1-phosphate). A minor form of galactosemia is caused by a deficiency in the enzyme during infancy.

UTP–xylose-1-phosphate uridylyltransferase

In addition to its other name, xylose-1-phosphate uridylyltransferase, UTP–xylose-1-phosphate uridylyltransferase is an enzyme that catalyzes a reaction between UTP and a-D-xylose 1-phosphate to generate UDPxylose and pyrophosphate.

Utrophin or Dystrophin-Related Protein

The protein utrophin or dystrophin-related protein (abbr.: DRP) is highly similar to dystrophin and is found in normal adult human muscle, particularly at the neuromuscular junction. This protein can be seen in the sarcolemma of patients with Duchenne muscular dystrophy who do not have dystrophin. It is an actin-binding protein that is similar to a-actinin in structure and function. The dystrophin gene and protein sequences are quite similar to those of dystrophin. Excessive expression of utrophin in a dystrophin-deficient (mdx) animal has demonstrated the ability of the protein to substitute dystrophin, raising the possibility of developing a treatment method for Duchenne muscular dystrophy.

It is a process in which a brief exposure (1 minute) to ultraviolet (UV) light (254 nm) enables RNA or DNA immobilized on nitrocellulose or nylon membranes to become covalently linked to the membrane, and it is used in genetic research.

DNA damage generated by ultraviolet light and DNA-reactive chemicals can be repaired by the SOS repair system, which is induced by four genes in *Escherichia coli.*

UVR

The products of these genes are involved in the repair of DNA damage caused by UV radiation and DNA-reactive chemicals. These three genes, designated by the letters uvrA, uvrB, and uvrC, encode the three subunits of the ABC excision nuclease. This enzyme recognizes the structural distortion caused by a thymine dimer and cleaves the damaged strand at two sites approximately 12 nucleotides apart, leaving an opening that is filled by polymerase and ligase action. Because the enzyme cuts at two different locations, it is not classified as a traditional endonuclease, but rather as an exconuclease. The fourth gene, uvrD, encodes a DNA helicase with ATPase activity, which is comparable to the product of the rep gene. It is involved in the ATP-dependent unwinding of the DNA duplex during DNA repair, and it is found in both humans and mice.

V

Vaccenatel

numerical symbol: 18:1(11); the trivial name for either of the isomers (cis-vaccenate and trans-vaccenate) of octadec-11- enoate, CH_3–$[CH_2]_5$–CH=CH–$[CH_2]_9$–coo$^-$, the anions derived from cis-vaccenic acid, (Z)-octadec-11-enoic acid, and transvaccenic acid, (E)-octadec-11-enoic acid, a stereoisomeric pair of monounsaturated straight-chain higher fatty acids.

Vaccenoyl

symbol: Vac; the trivial name for either of the isomers (cis-vaccenoyl and trans-vaccenoyl) of octadec-11-enoyl, CH_3–$[CH_2]_5$–CH=CH–$[CH_2]_9$–CO^-, the acyl groups derived from cis-vaccenic acid, (Z)-octadec-11-enoic acid, and trans-vaccenic acid, (E)-octadec-11-enoic acid. The cis-vaccenoyl group is a major fatty-acyl component of many bacterial lipids and a constituent of the oils of marine organisms. The trans-vaccenoyl group is present in a minor proportion of the acylglycerols in the body and milk fats of ruminants.

Vaccine

Any preparation of immunogenic material suitable for the stimulation of active immunity in animals without causing disease is referred to as a vaccination. Dead or attenuated microbes, altered toxins (toxoids), or viruses can all be used in the development of vaccines.

Vaccinia Virus

It is the type species of the Orthopoxvirus genus (also known as the 'vaccinia subgroup'), which comprises the viruses that cause cowpox, buffalopox, catpox, mousepox, and (formerly) smallpox. Vaccinia virus is a DNA virus and the type species of the genus Orthopoxvirus (or 'vaccinia subgroup') (variola virus). In most cases, a vaccinia infection produces only moderate symptoms. This virus was previously employed as the basis of a live vaccination to protect humans against smallpox, due to its serological similarity to the virus that causes the disease. Engineering vaccinias with surface antigens of hepatitis B virus, influenza viral surface antigens (which kill cattle, horses, and pigs), and vesicular stomatitis virus (which kills cattle, horses, and pigs) has resulted in vaccinations that have been effective in animal experiments. However, adverse effects to vaccinia-virus vaccines have been widely reported.

Vacuolar Apparatus

The vacuolar apparatus is a cellular digesting system that is composed of lysosomes and closely similar, hydrolase-free, vacuolar structures that are hydrolase-free. He or she is responsible for the digestion of both endogenous and exogenic elements.

Vacuole

A vacuole is a closed structure found solely in eukaryotic cells that is completely enclosed by a unit membrane and contains liquid material. It is defined as follows: Plant cells are distinguished by the presence of a prominent vacuole, which can take up to 95 percent of the total cell volume. Because it contains a high concentration of hydrolytic enzymes, it serves a function in these cells that is comparable to that of lysosomes in animal cells (especially various types of protease). In seed reserve tissues, specialized protein storage vacuoles store an abundance of storage protein that has been deposited during the process of seed development. When a seedling germinates, a specific protease is transported into the plant to provide amino acids to the new seedling.

Vacuum Evaporation

Vacuum evaporation is a process of depositing a thin film of a solid substance on a surface by evaporating the material at a high temperature in a vacuum while the material is still liquid. In order to produce a coating on the cool surface, atoms leaving the heated material travel directly to the cool surface without clashing with other molecules in the gas phase; they condense on the cool surface and accumulate as a layer of solid.

Valinomycin

It is composed of three moieties, each of which contains one molecule of each of the amino acids L-valine, D-a hydroxyvaleric acid, D-valine, and L-lactic acid. Valinomycin is a cyclic 12-residue depsipeptide antibiotic. The D-valine carboxyl group is immediately attached to the a-carbon of L-lactic acid through the a-carbon bonding system. It is made by the bacteria *Streptomyces fulvissimus* and has anti-tuberculosis activity, particularly against Mycobacterium tuberculosis. When it folds, it creates a hydrophobic surface with an interior that binds ions such as Rb+, Cs+, NH4 +, or, most importantly, K+. It interferes with oxidative phosphorylation by making the mitochondrial membrane permeable to K+ ions, which is detrimental to the process.

Van den Bergh Reaction

The van den Bergh reaction is a method for determining bilirubin that is based on the creation of a red azo dye by bilirubin and diazotized sulfanilic acid, which is used to determine bilirubin levels. In aqueous solution, bilirubin glucuronide produces a direct reaction, while in alcoholic solution, free bilirubin produces an indirect reaction, both of which are beneficial.

Van der Waals Force

The van der Waals force is a long-range force between molecules or sub-molecular groups that is effective from molecular spacings of more than 50 nm to interatomic distances. Such forces are always there, but they are weak, and they frequently contend with electrostatic forces for control of the environment. As a general rule, they do not follow a straightforward power law; for example, the force may be attracting at large separations but repulsive at tiny separations, and vice versa. Because the van der Waals free energy between two molecules or sub-molecular groups is largely dependent on their mutual orientation, these forces seek to align two molecules in such a way that their free energy of interaction is minimized.

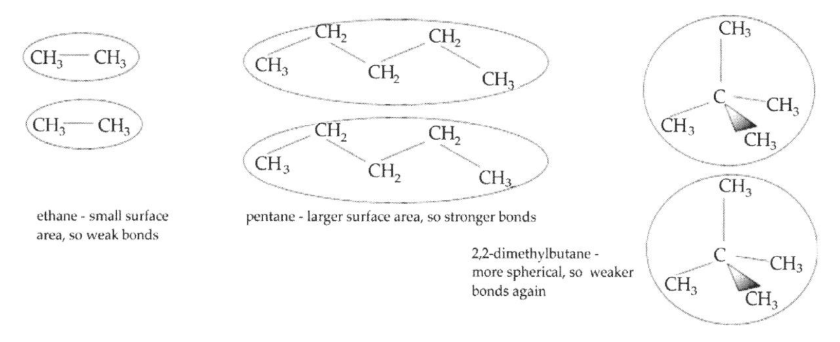

Figure 42. Alkane Strength of Van der Waals.

Source: Image by Wikimedia Commons

Van der Waals Radius

When the long-range attractive forces balance the short-range repulsive forces, the van der Waals radius is one-half of the internuclear distance between atoms at equilibrium. Unlike the covalent radius of an atom, the ionic radius of monovalent ions is almost equal to the radius of the atom's covalent radius.

Van't Hoff Law

It is the principle of Van't Hoff law that the osmotic pressure exerted by a solute is equal to the pressure that the solute would exert if it were an ideal gas at the same temperature as the solution and having the same volume as the solution As a result, $PV = RT$, where P denotes the osmotic pressure, V is the volume of solution containing one mole of solute, R denotes the gas constant, and T denotes the thermodynamic temperature

Variable

1. Having a wide range of possible values; being capable of deviating from a previously determined type
2. Something that is subject to change; a quantity or function that can take on any of a range of values that have been specified; a symbol that represents any unspecified number or quantity Variable arm is a term used to refer to the additional arm of the tRNA molecule.

Vasostatin

vasostatin is one of two segments of the chromogranin A molecule. Vasostatin I, which contains residues 1–76, lowers vascular stress, inhibits the release of parathyroid hormone, and has antifungal and bacteriostatic properties. Vasostatin II consists of residues 1–115 and shares the first two qualities of vasostatin I. Vasostatin II is composed of residues 1–115.

VASP

It is abbreviated VASP, which stands for vasodilator-stimulated phosphoprotein; it is a substrate for the platelet cyclic AMP- and cyclic GMP-dependent protein kinases. Heparin is found in association with actin filaments and focal adhesions, and it contains an EVH1 homology sequence that interacts with a proline-rich motif.

Vector Database

UniVec is an example of a vector database, which is a library of known vector sequences that may be utilized by vector-screening tools to discover sequence contaminants.

Vectorette PCR

In the field of polymerase chain reaction (PCR), vectorette PCR is a method for amplifying an unknown DNA sequence adjacent to a known DNA sequence using the polymerase chain reaction (PCR). It was formerly employed in the process of chromosome walking, but it continues to be useful in the recovery of sequences proximal to mutations induced by insertional mutations. In vectorial processing, a double-stranded vectorette is ligated to the ends of restriction fragments, thereby introducing a known sequence that can be used to prime one side of a PCR reaction, with the other side being primed on the known genomic sequence.

Vectorial Processing

Vectorial processing is the conversion of a fully formed precursor polypeptide into a mature protein when it is used to drive the movement of the protein from one aqueous compartment (e.g., the cytoplasm) through (e.g. an intramitochondrial or intrachloroplastal space).

Vectorial Translation

a stage of translation in the biosynthesis of any protein that is intended for export from a cell, during which a nascent polypeptide produced on a polysome bound to the membrane of a eukaryotic rough endoplasmic reticulum is conducted through the membrane by the N-terminal signal peptide of the polypeptide.

Vegetative Reproduction

1. Asexual reproduction in plants occurs when specialized multicellular organs (e.g., bulbs, corms, and tubers) break away from the parent plant and mature into new individuals.
2. In animals such as coelenterates and sponges, budding and other kinds of asexual reproduction are referred to as asexual reproduction (or asexual reproduction).

Vehicle

1. In order to increase the bulk of a biologically active material (such as a medicine) and/or to enhance its absorption after administration, it is necessary to mix it with an inert media, dissolve it, suspend it, or emulsify it in an inert medium.
2. any agent of transmission (animate or inanimate), particularly one that causes infection.

Venombin

In snake venom, venombin is one of two serine proteinases that can be detected. The enzyme venombin AB (EC 3.4.21.55), also known as gabonase (after the venom of the gaboon snake, *Bitis gabonica*), breaks down the Arg-|-Xaa bonds in fibrinogen to create fibrin and release the fibrinopeptides A and B. It is not inhibited by antithrombin III or hirudin. Venombin A has the EC number 3.4.21.74 and cleaves the Arg-|- peptide. In fibrinogen, Xaa forms fibrin and releases fibrinopeptide A, although the subsequent reaction is dependent on the species; it is a trypsin-like enzyme that bonds to fibrinogen.

Veratridine

Veratridine is a poisonous alkaloid with the chemical formula C36H51NO11 that can be found in the seeds of *Schoenocaulon officinalis* (of the lily family) and the rhizome of *Veratrum album*, among other plants. It interferes with the generation of the action potential that is involved in axonal conduction by binding to the Na+ channels and blocking them in the open state of the channel.

Versican

Versican, also known as big fibroblast proteoglycan or chondroitin sulfate proteoglycan core protein 2, is a protein that plays a role in intercellular communication. The N-terminal portion is similar to that of a glial hyaluronate-binding protein; the middle portion contains glucosaminoglucan attachment sites; and the C-terminal portion has EGF-like repeats and a lectin-like domain, among other features.

Very-Low-Density Lipoprotein

VLDL is an abbreviation for very-low-density lipoprotein, which is a type of lipoprotein found in the blood plasma of many animals (data normally relate to humans). VLDL is also referred to as pre-beta lipoprotein due to the fact that its electrophoretic mobility is somewhat greater than that of low-density lipoprotein, also known as beta-lipoprotein (LDL). VLDL particles have a diameter of 25–70 nm and a solvent density of (g mL–1) for isolation.

Very-Low-Density-Lipoprotein Receptor

VLDL receptor (an acronym for very-low-density lipoprotein receptor) is a membrane protein that binds to VLDL, accumulates in clathrin-coated pits, and internalizes VLDL in the body. This protein has a large extracellular domain, a single transmembrane domain, and a tiny intracellular domain, among other characteristics. The 873 amino acids that make up the human precursor protein are divided into three domains: the signal, the extracellular domain, and the cytoplasmic domain. The signal is composed of residues 1–27, the extracellular domain is composed of residues 28–797, and the cytoplasmic domain is composed of residues 820–873.

Viability

Viability is a measure of a cell's ability to perform metabolic functions and reproduce itself. It is frequently estimated experimentally by counting the number of cells that reject dyes that are ordinarily membrane-impermeant, such as Trypan Blue, from the sample.

Vicinal

1. The adjective vicinal means "nearby" or "neighboring."
2. A term used in organic chemistry to describe two (typically identical) atoms or groups that are bonded to two linked carbon atoms in a molecule, one to each of the two carbon atoms. The presence of vicinal substituents in a compound is indicated by the prefix vic- being appended to the name of the molecule's structure.

Vinculin

Vinculin is a cytoskeletal protein that is associated with the cytoplasmic face of focal adhesion plaques. It is responsible for anchoring actin microfilaments to the plasma membrane and for attaching a cell to the substratum. It works in conjunction with talin to bind integrins to the cytoskeleton of the cell. Vinculin is phosphorylated (on the amino acids serines, threonines, and tyrosines) and acylated by myristic acid and/or palmitic acid, and it is found in a variety of foods. Using alternative splicing of the same gene, the proteins vinculin and metavinculin are generated. Metavinculin varies from vinculin in that it contains an extra domain towards the C terminus that contains 68 residues.

Viscosity

symbol: g; the protection from the stream of a liquid because of the amount of the impacts of grip and attachment. Consider a fluid between two equal plates, one of which is moving in the x bearing with a speed v. The fluid is considered various layers, every one of which slides along the neighboring layer; the frictional obstruction between contiguous layers creates a speed angle in the y bearing. The misshapening of the fluid created by the speed inclination is known as shear.

Visualization Software

1. any product that works with investigation and translation of information utilizing straightforward visual similitudes (for example diagrams, trees, 3D pictures).
2. (in structure examination) programming that permits perception of 3D portrayals of molecular structures according to different points of view, utilizing numerous modes (for example spine follow, space filled, surface) and different shading plans; models incorporate MAGE, RasMol, and Cn_3D.
3. (in sequence analysis) software that allows visualization of sequence properties (such as hydropathy profiles) or visualization and editing of sequence alignments: e.g. CINEMA (color interactive editor for multiple alignments) and JalView.

Vitamin-Like Dietary Factor

Any biomolecule that looks like the nutrients in its organic properties and is viewed as delivered in lacking sums inside a organism with the goal that dietary supplementation might be suggested. The term includes: inositol, carnitine, lipoic acid, p-aminobenzoic acid, and ubiquinone.

Vitellogenin

The forerunner protein of lipovitellin and phosphitin, a significant egg-yolk protein combined in the liver of chickens and Xenopus. It has a Mr of 135 000 and is split in the ovary into its constituent proteins. In Xenopus, two genes, for vitellogenin I and II, are spread over more than 21 kbp of DNA, their essential parts comprising of 6 kbp of pre-mRNA with 33 introns. A connected protein is found in *Drosophila melanogaster*.

Volkensin

A ricin-like harmful glycoprotein from the underlying foundations of *Adenia volkensii*, a Kenyan bush. It comprises two subunits, A (Mr 29 000) and B (Mr 36 000), connected by disulfide bonds.

Waardenburg Syndrome

An autosomal predominant disorder of formative deformities in and around the eyes, pigmentary deserts in eyes and hair, and intrinsic deafness. Types 1 and 3 are brought about by transformations in PAX-1 and PAX-3. Type 3 likewise shows appendage muscle hypoplasia. Type 2 is brought about by transformations in MITF; type 4 is related to Hirschsprung sickness, and is brought about by transformations in SOX10.

Walker Motif

Both of two arrangement themes in proteins that tough situation and hydrolyze nucleoside triphosphates. Walker-A (or P-circle) comprises of the arrangement A/GX4GKT/S (in one-letter code), is flanked by a beta-strand and an alpha helix, and circles around the triphosphate moiety. Walker-B comprises X4D (where X is nearly solely a hydrophobic residue), happens toward the terminal of a beta-strand, and collaborates with the Mg 2+ particle of the triphosphate moiety.

Wall Effect

1. (in centrifugation) the impact of sedimenting molecules with the mass of the cell. This happens in light of the fact that the divergent field is outspread, so can be kept away from by utilizing area formed cells.
2. (in chromatography) the bending and spreading of a zone as it moves down a chromatographic section as a result of inhomogeneities in a dissolvable stream close to the mass of the segment.

Warburg Apparatus

A delicate, consistent volume respirometer for estimating gas trade of cells, homogenates, or tissue cuts. It comprises of a little jar, with at least one side arms for the expansion of reagents, and a middle all things considered, associated with a U-tube manometer of around 1 mm 2 inner cross-segment, fitted with a repository for manometer liquid. The vessel is drenched in a consistent temperature shower and shaken ceaselessly to equilibrate the gas in the fluid stage, where the response is occurring, with the gas stage, where the response is being estimated. It was utilized broadly for metabolic investigations before the wide availabilty of radioactive tracers, however, is currently basically outdated.

Waring Blender

The exclusive name for a blender utilized in the readiness of tissue homogenates. It comprises a hydrodynamically planned vessel (of tetrafoil cross-area) in the lower part of which an exceptionally formed cutting edge turns at high velocity.

WASP

abbr. for Wiskott-Aldrich disorder protein; any of a little group of proteins including WASP1 (whose expression is restricted to lymphocytes and platelets) and N-WASP (which is all the more generally expressed however, particularly in neurons). They bind little GTPases and phospholipids through their N-terminal areas and are homologous in their C-terminal districts. They act on the Arp2/3 complex and help in actin fiber nucleation. Around 20 changes of a locus at Xp11.23-p11.22 for WASP (501 amino acids) are related with the Wiskott-Aldrich condition of extreme thrombocytopenia and prolonged bleeding.

Wasserman Reaction

A complement-fixation test for human blood and cerebrospinal fluid, widely used to detect syphilitic infection, although false-positive reactions are often given by yaws, leprosy, paroxysmal hemoglobinuria, and malarial infection. The antigen on which the test is based is cardiolipin.

Watanabe Rabbit

A variety of rabbits with high blood cholesterol levels because of change of the low-density lipoprotein (LDL) receptor quality. It is utilized as an animal model for familial hypercholesterolemia.

Watson–Crick Model of DNA

A model for DNA comprising basically of two antiparallel helical polynucleotide strands coiled around a frame as a twofold helix. The deoxyribose-phosphate backbones lie on the outside of the helix and the purine and pyrimidine bases lie around at right points to the node on within the helix. The measurement of the helix is 2.0 nm and there is a residue on each chain each 0.34 nm in the z chain. The angle between every residue on a similar strand is 36°, with the goal that the design repeats after 10 residues (3.4 nm) on each strand. The two strands are kept intact by hydrogen bonds between sets of bases that are corresponding and complementary (for example on various strands). Adenine bonds with thymine and guanine with cytosine. The two chains are accordingly correlative. Watson and Cramp's paper in Nature in 1953, likewise contained a judicious proclamation: 'It has not gotten away our notice that the particular blending we have hypothesized promptly recommends a potential replicating component for the hereditary material.' This structure has endured over the extreme long haul. It is challenging to overstate the significance of the short paper. To cite from L. Stryer's Biochemistry (fourth release), 'This splendid achievement positions as one of the most critical throughout the entire existence of science since it drove the best approach to an comprehension of quality capacity in sub-molecular terms.'

Watson Strand

The upper strand in the double-stranded DNA coils. Open reading frames (ORFs) may occur on one or the other strand of a genomic DNA duplex and it is helpful to recognize the strands to indicate specific ORFs. ORFs in the

Watson strand of the yeast genome, for instance, are given the addition W, as in the official name YLR451W for the LEU3 gene on yeast chromosome XII.

Weak Interactions

Noncovalent interactions between molecules and parts of molecules, including charge–charge interactions, those involving permanent dipoles, van der Waals forces, and hydrogen bonds. They are particularly important in protein and nucleic-acid structures.

Weiss Unit

A proportion of T4 DNA ligase action in view of ATP-PPi exchange. One unit is how much compound expected to catalyze the exchange of one nanomole of 32P from 32PPi into ATP as Noritadsorbable material in 20 min at 37°C. ATP and 32P labeled ATP are adsorbed by Norit and recovered by filtration in this state for counting of radioactivity.

Werner Syndrome

An interesting autosomal recessive issue described by untimely arrangement of wrinkles, scleroderma-like skin, short height, turning gray hair and going bald, a summed up appearance of untimely aging, and a high risk of harm. It is brought about by mutations in the Werner disorder gene.

Wernicke–Korsakoff

A condition with a mix of Wernicke's disease (ophthalmoplegia, ataxia, and confusion) and Korsakoff's psychosis (trouble in recording and holding new experiences). Wernicke's disease is because of lack of vitamin B1, and shows especially in patients with a hereditary mutation in transketolase bringing about its decreased affinity for thiamine diphosphate. In such patients, even mild B 1 deficiency, as frequently happens in heavy drinkers, may create indications.

Western Blotting

A strategy for blotting proteins onto nitrocellulose, nylon, or another exchange layer after they have been resolved by gel electrophoresis. The proteins can then be recognized via autoradiography (whenever marked), or then again through restricting to fluorescently named, 125 I-marked, or chemical connected neutralizer, lectin, or other specific molecules. The name infers likewise through the cardinal applications from Southern blotching, the first such smudging procedure, with a capital starting by comparative expansion.

White Adipose Tissue

Or on the other hand white fat that is a tissue especially in well adapted animals and birds to store fat as triacylglycerols to supply energy to the entire organism. The cells contain a solitary enormous drop of fat that, in the fed state, fills the greater part of the cell. The nucleus lies in the thin band of cytoplasm around the fringe. Fat tissue is found in many pieces of the body yet is concentrated under the skin (subcutaneous) and around inside organs (heart and kidney).

Whole Genome PCR

A strategy by which genomic DNA is changed over to a structure that can be intensified by the polymerase chain response (PCR). DNA is decreased to little parts by limiting protein processing or sonication, and adaptors are ligated to their ends. Oligonucleotides specific for the adaptors can be utilized for PCR amplification. The strategy can be utilized to distinguish individual parts prepared to do restricting digestion or other DNA-restricting proteins involving antibodies specific for the protein concerned.

Williams–Beuren Syndrome

An autosomal dominant gene deletion including 17 genes on chromosome 7 and described by a wide range of anomalies. The gene at 7q11.23 for FKBP6 (a part of the synaptonemal complex) is much of the time part of the deletion.

Wilson's Disease

Hepatolenticular degeneration is an interesting autosomal recessive acquired disease described by degenerative changes in the cerebrum, especially in the basal ganglia, and cirrhosis of the liver. There are excess stores of copper in the liver, cerebrum, cornea, and kidney; the serum copper is normally low and there is a lack of ceruloplasmin. Patients might be actually treated with British antilewisite (dimercaprol) or other chelating specialists.

Wiskott–Aldrich Syndrome

A disorder of thrombocytopenia, dermatitis, and immunodeficiency, with higher chance of threat. It is brought about by mutations in a protein encoded on the X chromosome.

WNT Family

A group of secreted glycoproteins (for example WNT-2 from human and mouse) that are flagging atoms in tissue aging. They act through G-protein-related membrane receptors of the frizzled family to repress glycogen synthase kinase 3. The WNT-1 quality is intently homologous to wingless.

WNT Inhibitory Factor 1

abbr.: WIF-1; a secreted protein that contains an N-terminal WIF domain (≈150 amino acids), which binds and represses Wnt proteins; five EGF-like repeats; and a short hydrophilic domain in the C-terminal. Homologs are present in numerous organisms. The WIF domain is connected with the extracellular domain of Ryk receptor tyrosine kinases.

Wobble Hypothesis

The speculation that in interpretation (during protein synthesis) a less severe match in the base matching of the 5′ base of the anticodon of move RNA (tRNA) permits it to make hydrogen-bonding with the third base of the codon of messenger RNA (mRNA) past the typical G-C and A-U pairings. In this manner, anticodon U might match or G and anticodon G might match C or U. Albeit the speculation has been completely proven by experiments, UGG and AUG are the main instances of the primary sort of wobble, what's more, no instances of the subsequent kind are known. The presence of bases other than U, A, G, or C at the anticodon 5′ position additionally adds to a more broad freedom at this position. Consequently, bases, for example, wyosine, queuosine, and inosine might be available (I generally fill in for A) with less severe matching requirements. The wobble at the third codon position is steady with the way that around 40 distinct tRNA molecules cooperate with the 61 potential sense codons.

Writhing Number

image: W; a list of the superhelical winding of DNA, named coiling. The coiling number and the bend number between them decide the connecting number (L). The writhing number doesn't have a exact quantitative definition, yet addresses the level of supercoiling. A decline in L includes some decline in supercoiling, and an expansion in L an increment in supercoiling.

WRNp

abbr. for Werner disorder gene protein; other name: RECQL3. A RecQ helicase (1432 amino acids) of proteins that have 3′→5′ exonuclease movement and bind to the ssDNA-restricting protein replication protein A. Over a score of mutations cause Werner condition.

WT33

A protein that is connected with early development proteins, and is related to Wilm's tumor; it is presumably a tumor silencer. Certain mutations are likewise connected with Denys-Drash disorder. Its structure contains DNA binding domains and a zinc finger.

WW domain

WWP domain is a described by two conssserved tryptophan residues (~20-23 amino acid residues apart) and a proline, thus its name. It is seen in different divergent proteins, including vertebrate YAP protein, mouse NEDD-4, utrophin and dystrophin, the protein encoded by the gene responsible for Duchenne strong dystrophy. The domain contains ~35-40 residues, which crease as a triple-antiparallelll b-sheet, and might be repeated a few (up to 4) times. It binds proteins that contain specific proline patterns ([AP]-P-P-[AP]-Y), and looks similar to SH3 domains. It is frequently connected with different domains associated with signal transduction.

Wyosine

W or Y; any adjusted type of guanosine present in the anticodon circle of certain types of transfer RNA. They depend on N 2 - (2-methyletheno) guanosine, and transport changing side chains on the five-carbon imidazo ring.

Xanthan Gum

A capsular complex heteropolysaccharide formed by strains of the pseudomonad bacterium *Xanthomonas campestris* and accepted to boost attachment of the organism to its plant host. It is broadly utilized in preservation, food, and different uses as a crystallization inhibitor, emulsifier, gelling specialist, and so on.

Xanthine

2,6-dihydroxypurine; 2,6-dioxopurine; 3,7-dihydro-1H-purine-2,6-dione; a purine formed in the metabolic breakdown of guanine.

Xanthine Oxidase

The suggested name for hypoxanthine oxidase; efficient name: xanthine:oxygen oxidoreductase; previous name: Schardinger catalyst. A flavoprotein (FAD) catalyst containing two 2Fe-2S groups and molybdenum cofactor, present in milk and liver, that catalyzes the oxidation of xanthine to urate and superoxide (or hydrogen peroxide) and furthermore the oxidation of hypoxanthine to xanthine. The catalyst has two structures: (1) xanthine dehydrogenase, which has NAD as a coenzyme, and urate and NADH as items; furthermore (2) xanthine oxidase, which responds with dioxygen and has urate and H2O2 as items. The dehydrogenase is changed over reversibly to the oxidase by oxidation of sulfhydryl groups or irreversibly by proteolysis.

Xanthinuria

The discharge of extreme levels of xanthine in the urine. It is at times because of a genetic disease in which there is a lack of gross xanthine oxidase transport in the tissues.

Xanthoma

A yellow greasy store beneath the skin or connected with a ligament. Typically different, xanthomas are expected to extravascular phagocytosis of chylomicrons, LDL- cholesterol, or different sterols, by macrophages of the skin or subcutaneous tissue. They as a rule happen in hyperlipidemias and in cerebrotendinous xanthomatosis and may likewise happen with an ordinary plasma lipid profile.

Xanthoproteic

A test for protein in which a yellow color forms on the reaction of conc. nitric acid; the color becomes orange with antacid.

Xanthosine Monophosphate

abbr.: XMP; any xanthosine phosphate, especially xanthosine 5′-phosphate when its differentiation from xanthosine (5′-)diphosphate and xanthosine (5′-)triphosphate requires accentuation.

Xanthosine Phosphate

Xanthosine monophosphate (abbr.: XMP); any phosphoric monoester or diester of xanthosine. Of the three potential monoesters and the two potential diesters, just xanthosine 5′-phosphate occurs normally (the locant being excluded assuming no uncertainity).

Xanthylic Acid

The trivial name for any phosphoric monoester of xanthosine. The place of the phosphoric residue on the ribose moiety of a given ester might be determined by a prefixed locant. In any case, 5′-xanthylic acid is the ester ordinarily indicated, its locant generally being discarded assuming no uncertainty might emerge.

X Chromosome

It is generally expected found matched in the homogametic sex, which in numerous species is the female, and single in the heterogametic sex, in numerous species the male. The X chromosome conveys countless genes that control development and capacity. A mutant quality so conveyed is called X linked. Since males have just a single X chromosome, an unusual gene that it conveys can't be matched with a typical allele; the male is then supposed to be hemizygous for that gene. The female with two indistinguishable individuals from a pair of X-connected genes is supposed to be homozygous for that quality; the female with one mutant quality is known as a heterozygote or transporter for the unusual allele. One X chromosome is inactivated during early advancement. Since the offspring of every cell has a similar inactivated X chromosome, an extent of cells have the fatherly X chromosome in a functioning state while in different cells the maternal X chromosome is dynamic. Henceforth the female who is heterozygous for a new quality will have cells with the typical traits and cells with the mutant quality. She is supposed to be a heterozygote.

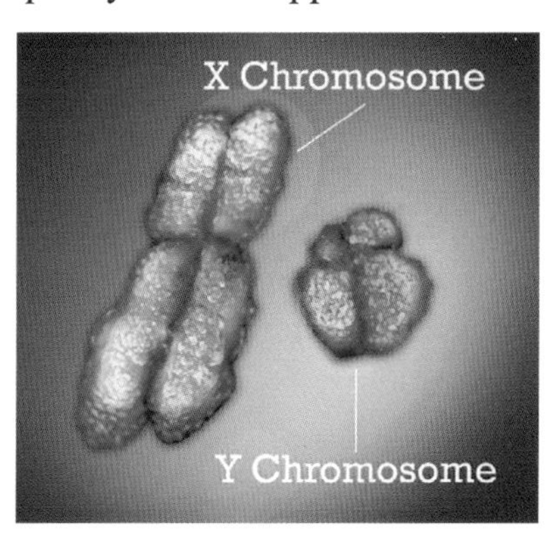

Figure 43. An Illustration of X Chromosome and Y Chromosome.

Source: Image by Flickr

X-Chromosome Inactivation

X inactivation is the inactivation in female warm blooded organisms of one of the two X chromosomes during development. This peculiarity happens haphazardly in cells of the female organism, which is consequently a heterozygote in regard to whether the maternal or fatherly X chromosome is functionalll. X inactivation implies that the measurement of genes in female cells is practically identical with that of males, who acquire just a single X chromosome.

Xenical (Orlistat)

A thinning drug that represses lipase in the stomach so that fat isn't consumed. Subsequently, around 30% of the fat in a food is indigested.

X-Ray Crystallography

The investigation of the mathematical types of crystals by X-beam diffraction. Myoglobin was the principal protein for which the construction was seen by X-beam crystallography.

X-Ray Diffraction

A technique that utilizes the diffraction design got by going X-beams through crystals, or other normal atomic clusters, to gauge interatomic distances and to decide the three-layered plan of atoms (or atoms) in the molecule. For the example to be sharp, the frequency of the radiation utilized should be more limited than the normal separation between the atoms of the structure. X-beams normally have frequencies of a couple of tenths of a nanometre, making them reasonable for organic molecules. X-beam diffraction is utilized broadly in the structures of proteins, nucleic acids, and other natural molecules (an early milestone was the assurance of the design of penicillin). To resolve the minor distinctions, much use has been made of isomorphous supplanting with heavy metal atoms.

X-Rays

Or then again (at times, esp. US) x-beams or (previously) röntgen beams electromagnetic radiation produced by particles during an extranuclear loss of energy of episode radiation (for example high-energy electrons, gamma radiation) or by atoms of certain radionuclides during change by electron catch. X-beams have frequencies in the reach 1 pm-10 nm (frequencies

3 PHz-3 EHz), which lies over that of gamma radiation and covers the frequencies of the far bright. Trademark X-beams have specific frequencies of a specific nuclide or target component.

Xylan

Any homopolysaccharide of xylose, comprising mostly of b-1,4-connected xylopyranose units with different substituent groups, that happens in the cell walls of higher plants, and in some marine algae. Xylans structure the major glycans of the hemicellulose in the strong parts of plants.

Xylem

The vascular tissue that behaviors water and mineral salts from the roots to the aerial pieces of a plant and gives it with mechanical help.

Xylose

A constituent of plant polysaccharides (see xylan); the normally occuring enantiomer is dependably D-xylose. In people, most ingested xylose is discharged unaltered in the urine. This is the reason for the xylose assimilation test, the most generally utilized test of starch retention.

D-xylulose 5-phosphate

The 5-phosphate ester of D-xylulose and a part of the pentose phosphate pathway. It is formed from Dribulose 5-phosphate by ribulose-phosphate 3-epimerase (EC 5.1.3.1); D-xylulose 5-phosphate then, at that point, reacts with D-ribose 5-phosphate, catalyzed by transketolase, to give glyceraldehyde 3-phosphate and sedoheptulose 7-phosphate.

Y chromosome

A sex chromosome found distinctly in the heterogametic sex and typically disparate in size from the X chromosome. In numerous organisms it conveys the testis-deciding element that triggers male early-stage development.

Yeast

Any of a group of unicellular growths that repeat agamically - by sprouting or parting - and physically – by the development of ascospores. Yeast cells might occur independently or in short chains, and a few animal varieties produce a mycelium. The term 'yeast' is frequently used to mean individuals from = Saccharomyces, for example, *S. cerevisiae*, which is an example of a maturing yeast, or *Schizosaccharomyces pombe*, which is an illustration of a budding yeast.

Yeast Artificial Chromosome

abbr.: YAC; a particular cloning vector that contains a centromere, an autonomously replicating system (ARS), a couple of telomeres, selectable marker genes, and the part of DNA to be cloned. As a rule, genomic DNA is processed to form pieces containing the genes, which are isolated by beat field gel electrophoresis; the enormous parts are then ligated into YACs. They are prepared to do engendering in yeasts, where they work as 'counterfeit chromosomes', being productively recreated. YACs are fundamental in enormous planning projects, for example, the Human Genome Project, as they are steady and can convey tremendous DNA regions of 100 to 1000 kbp.

Yeast Cloning

A technique for cloning involving yeast as the host. Articulation of eukaryote genes by yeasts and other microbial eukaryotes is dependent upon eukaryotic administrative systems, and the items go through post-translational modifications as opposed to when they are expressed in prokaryotes. Accordingly, yeast clones are frequently utilized for the expression of glycoproteins.

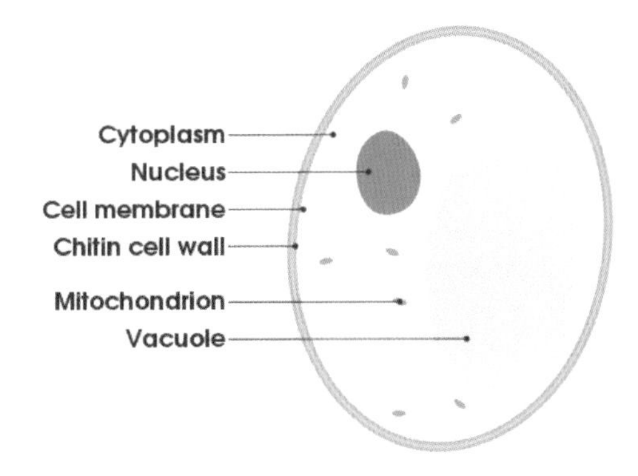

Figure 44. Simple Diagram of Yeast Cell.

Source: Image by Wikimedia Commons

Yeast Two-Hybrid System

A system that takes advantage of two crossover proteins, X, intertwined to the GAL4 DNA-binding domain, and Y, combined to the GAL4 initiation domain. Plasmids encoding every one of the crossover proteins are brought together into yeast, which prompts the two crossovers. The system is utilized to test whether protein Y binds to protein X, for instance in exploring the job of new proteins. Assuming the proteins do bind, the coming about complex will be bound to the upstream initiation arrangement for the yeast GAL genes by the GAL4 DNA-binding domain, and the presence of the GAL4 domain will prompt transcriptional initiation of the reporter gene, for example, b-galactosidase. The system enjoys the upper hand over immunoprecipitation that assuming Y binds to X, the quality encoding it is now accessible as a clone. The half-breed containing X is alluded to as the trap.

Yellow Enzyme

1. old yellow enzyme, EC 1.6.99.1; NADPH dehydrogenase; a flavo-protein (FMN in yeast, FAD in plants).
2. new yellow enzyme an alternative name for D-amino-acid oxidase.

Yellow Fever Virus

There are two sorts of fever; wilderness which is a disease of monkeys that is spread by contaminated mosquitoes; uncommon in people; and metropolitan, an illness of people brought about by contaminated mosquitoes *Aedes aegypti*. The infection is an individual from the Flaviviridae with a sense in addition to RNA genome meant a polyprotein of envelope, layer and capsid proteins. Inoculation is by a live-weakened 17D strain.

V-YES

It is the oncogene of the Yamaguchi sarcoma, and has a cell partner, c-indeed, the site of which is a member from the src group of non-receptor tyrosine kinases, having a myristoylation site, and SH2, SH3 and tyrosine kinase areas.

Yield

Or Y; a proportion expressing the proficiency of a mass transformation process. The yield coefficient is characterized as how much cell mass or product formed connected with the consumed substrate or to the intracellular ATP formation.

Yin-Yang Hypothesis

A model of natural guidelines in light of the contradicting activities of specific cyclic nucleotides. It emerges from the perception that chemicals or other naturally dynamic substances that boost the cell aggregation of cyclic

GMP produce cell reactions that are opposing to ones happening when the synthesis of cyclic AMP is expanded in similar tissues or cells. Henceforth cyclic GMP and cyclic AMP may have restricting or opposing control impacts in specific organic systems, by similarity with the oriental idea of yin and yang representing a dualism between contradicting normal powers.

+yl

In compound terminology, indicating a free valence got from the departure of a hydrogen atom from a parent hydride, for example, methyl, pentanyl. It additionally applies to specific acyl groups, which are exemptions for the standard that acyl groupes are named by changing the '- ic acid' or '- oic acid' finishing to '- oyl'. These exemptions are acetyl, malonyl, succinyl, propionyl, butyryl, oxalyl, and glutaryl.

Yohimbine

An indole alkaloid, $C_{21}H_{26}N_2O_3$, with α^2 - adrenoceptor antagonist function. It is sourced from *Corynanthe johimbe* and *Rauwolfia serpentina.*

YEP

abbr. for Yersinia external protein; any of various proteins, known as pathogenic proteins, initially seen as on the outer layer of Gram-negative microbes ofYersinia; they are presently known to be normal for plant-pathogenic proteins that are released by microorganisms of genera as different as Pseudomonas, Xanthomonas, and Erwinia. They are additionally released by microorganisms, like *Shigella flexneri*, *Salmonella typhimurium*, and *Escherichia coli*. They are released by the type III system.

Z

ZAP

abbr. for zinc finger antiviral protein; a protein (776 amino acids) of rodents that contains four zinc fingers (of C_4H type). At the point when expressed in fibroblasts in cell culture it presents specific antiviral functions.

Zebrafish

Brachydanio rerio; a little, exceptionally rich type of characinoid teleost fish with a sparkly blue body and four longitudinal yellowish stripes along its sides. It is quite utilized in investigations on the molecular science of separation and improvement in metazoan animals.

Zeeman Effect

The splitting of a solitary line in a range, demonstrative of the degeneration of the energized condition of a specific chromophore, into at least two parts of marginally various frequencies achieved by the use of an outer magnetic field

Zein

The foremost storage protein of maize (corn) seeds. It is moderately lacking in the fundamental amino acids lysine and tryptophan; tryptophan inadequacy can occur in human populaces subject to maize as the main protein source. The zein genes are grouped.

Zellweger Syndrome

Or then again cerebrohepatorenal condition a heterogeneous problem of peroxisome biogenesis that outcomes from mutations in 11 loci for PEX proteins, most normally for PEX1. Commonly patients have various innate irregularities of the face and head, progressing metabolic conditions with aggregation of very- long-chain unsaturated fats, and for the most part die in early stages. Childish Refsum's sickness is one type of this condition.

Zeolite

Any permeable soluble base metal-or antacid earth-aluminum silicate that shows particle exchange properties. Zeolites might be utilized for water softening or as sub-molecular filters.

Zeta Chain

1. A transmembrane protein (16 kDa) of most T cells that contains three intracellular immunoreceptor tyrosine-based activation motifs (ITAMs) to which protein tyrosine kinases binddd. It forms a dimer of disulfide-connected chains that are firmly connected with parts of CD3 and T-cell receptors.
2. the a-globin-like globin that with c-globin structures Hb Portland I

Zeugmatography

A procedure where the expansion of painstakingly controlled inhomogeneous magnetic fields empowers estimations of nuclear magnetic resonance (NMR) to be made subsequently the arrangement molecules or practically any property quantifiable by NMR.

Zimm–Crothers Viscometer

A changed Couette viscometer in which the internal chamber is a self-centring float containing a steel pellet. The inward chamber is turned by an outside pivoting magnetic field, the speed of revolution being reliant, bury alia, on the thickness of the liquid. The instrument just measures the viscosiity of a testcomparative with a standard.

Zimmermann Reaction

The response of m-nitrobenzene in firmly basic preparation with the methylene group at position 16 of every 17-ketosteroids to give a purple color with an absorbance maximum at 520 nm. It very well might be utilized to gauge such steroids.

ZIP

1. abbr. for the leucine zipper area.
2. a group of membrane proteins engaged with the import of zinc into the cytoplasm in eukaryotes. They are anticipated to contain eight transmembrane sections, of which S4 and S8 are profoundly moderated and contain fixed histidine residues, which might be engaged with metal binding.

Zollinger–Ellison Syndrome

A condition brought about by inordinate secretion of gastrin, either from (type 1) G-cell hyperplasia in the antrum of the stomach, or from (type 2) a harmless or dangerous pancreatic islet cancer (gastrinoma). The signs incorporate various peptic ulcers, or peptic ulcers in uncommon locations, with an obvious propensity to dying, regularly connected with hyperparathyroidism.

Zone Electrophoresis

A technique for electrophoresis where the protein (or other) preparation is put at the as a small band or spot in an inert supporting medium (paper, starch gel, polyacrylamide gel, and so on). An electric potential is then applied to the supporting medium, causing the proteins (or different substances) to relocate to give distinct groups or zones. These might be located in situ by staining, light retention, and so forth, or by examination after elution of the bands of the supporting medium.

Zone Precipitation

A procedure where protein is accelerated as a zone in a gel-filtration segment by eluting with a gradient of a protein targeting molecule.

Zone Spreading

The expanding of a zone in electrophoresis or chromatography because of current flows or different unsettling influences in the supporting medium.

Zoo Blot

A strategy for recognizing the conservation of DNA preparation during tests. A section of DNA being examined is utilized as a test to hybridize against a progression of DNA tests from different species, and subsequently lay out whether the DNA succession has been conserved during development.

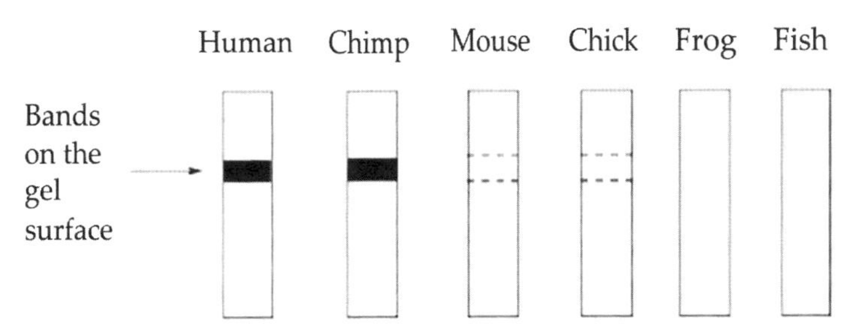

Figure 45. An Illustration of Zoo Blot.

Source: Image by Wikimedia Commons

Zymase

The name initially given to the heat labile, non - diffusible part of an unrefined concentrate of brewers' yeast that, with the expansion of the heat steady, diffusible portion (for example cozymase), would empower the heavy alcohol fermentation of glucose to happen without a cell system. It was subsequently displayed to comprise of a combination of chemicals, including those of the glycolytic pathway.

Zymogen

or on the other hand proenzyme or (previously) proferment the dormant antecedent of a compound, regularly convertible to the chemical by halfway proteolysis. The term is applied particularly to chemically inert types of pancreatic chemicals like trypsinogen, chymotrypsinogen, proelastase, and procarboxypeptidase; these are cleaved to let the active protein after their secretion out of granules (zymogen granules) in the acinar cells of the pancreas.

Zymogen Granule

A membrane bound, cytoplasmic secretory granule noticeable by light microscopy. Zymogen granules are framed in the Golgi apparatus of compound discharging cells and store a zymogen. The term is utilized particularly of a secretory granule containing the proenzyme of a stomach-related protein.

Zymogram

1. any record of a zone electrophoresis division where compounds in an example have been isolated and their positions, and relative sums, uncovered by an action stain.
2. any table appearance the aftereffects of sugar maturation tests completed during the time spent recognizing a microorganism.

Zyxina

The protein part of cell foundation and cell-cell adherens intersections. It is a part of attachment plaques and the ends of stress fibers; it collaborates with α-actinin.

REFERENCES

1. Biochemistry (Moscow), 2001. [online] 66(7), pp.818-818. Available at: <https://link.springer.com/article/10.1023/A:1017381200722> [Accessed 16 March 2022].
2. CATTELL, K., 1976. Dictionary of Biochemistry. Biochemical Society Transactions, [online] 4(5), pp.937-937. Available at: <https://portlandpress.com/biochemsoctrans/article-abstract/4/5/937/60934/Dictionary-of-Biochemistry?redirectedFrom=fulltext> [Accessed 16 March 2022].
3. Howe, C., 1990. Dictionary of Biochemistry and Molecular Biology (2nd edn). Trends in Genetics, [online] 6, p.228. Available at: <https://www.sciencedirect.com/science/article/abs/pii/016895259090184 8?via%3Dihub> [Accessed 16 March 2022].
4. Howe, C., 1998. Erratum: Oxford Dictionary of Biochemistry and Molecular Biology. Trends in Biochemical Sciences, [online] 23(12), p.468. Available at: <https://www.cell.com/trends/biochemical-sciences/fulltext/S0968-0004(98)01321-8?_returnURL=https%3A%2F%2Flinkinghub.elsevier.com%2Fretrieve%2Fpii%2FS0968000498013218%3Fshowall%3Dtrue> [Accessed 16 March 2022].
5. Reference Reviews, 2000. Oxford Dictionary of Biochemistry and Molecular Biology. [online] 14(8), pp.26-26. Available at: <https://www.emerald.com/insight/content/doi/10.1108/rr.2000.14.8.26.404/full/html> [Accessed 16 March 2022].
6. Šesták, Z., 2000. Oxford Dictionary of Biochemistry and Molecular Biology. Revised Edition. Photosynthetica, [online] 38(4), pp.606-606. Available at: <http://ps.ueb.cas.cz/artkey/phs-200004-0023_oxford-

dictionary-of-biochemistry-and-molecular-biology-revised-edition.php> [Accessed 16 March 2022].

7. The International Journal of Biochemistry & Cell Biology, 1998. Oxford dictionary of biochemistry and molecular biology. [online] 30(11), pp.1279-1280. Available at: <https://www.sciencedirect.com/science/article/abs/pii/S1357272598000387?via%3Dihub> [Accessed 16 March 2022].
8. Wood, E., 1997. The oxford dictionary of biochemistry and molecular biology. Biochemical Education, [online] 25(4), p.251. Available at: <https://www.sciencedirect.com/science/article/abs/pii/S0307441297875589?via%3Dihub> [Accessed 16 March 2022].
9. Wood, E., 2001. Tachenwörterbuch der biochemie/pocket dictionary of biochemistry, P. Reuter; Birkhäuser, Basel, 2000, SFr58, 560 pp., ISBN 3-7643-6197-2. Biochemistry and Molecular Biology Education, [online] 29(2), pp.89-90. Available at: <https://www.sciencedirect.com/science/article/abs/pii/S1470817500000485?via%3Dihub> [Accessed 16 March 2022].
10. Wood, E., 2007. The Oxford dictionary of biochemistry and molecular biology (second edition). Biochemistry and Molecular Biology Education, [online] 35(4), pp.311-311. Available at: <https://iubmb.onlinelibrary.wiley.com/doi/10.1002/bmb.70> [Accessed 16 March 2022].
11. Bandyopadhyay, P., Das, N. and Chattopadhyay, A., 2022. Biochemistry. Biochemical, Immunological and Epidemiological Analysis of Parasitic Diseases, [online] pp.245-261. Available at: <https://link.springer.com/chapter/10.1007/978-981-16-4384-2_6> [Accessed 16 March 2022].
12. Datta, S., 1976. Dictionary of biochemistry. FEBS Letters, [online] 67(2), pp.234-235. Available at: <https://febs.onlinelibrary.wiley.com/doi/abs/10.1016/0014-5793%2876%2980386-9> [Accessed 16 March 2022].
13. Garner, R., 1944. Dictionary of biochemistry and related subjects, Editor-in-Chief, William Marias Malisoff, Professor of Biochemistry at the Polytechnic Institute of Brooklyn. Philosophical Library, Inc., New York. 579pp. Price, $7.50. Journal of the American Pharmaceutical Association (Scientific ed.), [online] 33(1), p.31. Available at: <https://www.sciencedirect.com/science/article/abs/pii/S0095955315307010?via%3Dihub> [Accessed 16 March 2022].

14. Gwatkin, R., 1990. Dictionary of biochemistry and molecular biology, by J. Stenesch, 2nd ed., John Wiley & Sons, New York, 1989, 525 pp, $59.95. Molecular Reproduction and Development, [online] 27(2), pp.180-180. Available at: <https://onlinelibrary.wiley.com/doi/10.1002/mrd.1080270216> [Accessed 16 March 2022].
15. Jakoby, W., 1998. Oxford Dictionary of Biochemistry and Molecular Biology. Edited by A. D. Smith, S. P. Datta, G. H. Smith, P. N. Campbell, R. Bentley, and H. A. McKenzie. Analytical Biochemistry, [online] 258(2), p.382. Available at: <https://www.sciencedirect.com/science/article/abs/pii/S000326979892574X?via%3Dihub> [Accessed 16 March 2022].
16. Johnson, G., 2007. Oxford Dictionary of Biochemistry and Molecular Biology (2nd edition)2007235Edited by Richard Cammack and others. Oxford Dictionary of Biochemistry and Molecular Biology (2nd edition). Oxford: Oxford University Press 2006. xv+720 pp., ISBN: 978 0 19 852917 0 £49.95 $89.50. Reference Reviews, [online] 21(5), pp.41-42. Available at: <https://www.emerald.com/insight/content/doi/10.1108/09504120710755635/full/html> [Accessed 16 March 2022].
17. Kilby, B., 1976. Dictionary of Biochemistry. Biochemical Education, [online] 4(2), p.38. Available at: <https://www.sciencedirect.com/science/article/abs/pii/0307441276900431?via%3Dihub> [Accessed 16 March 2022].
18. Perlès, R., 1976. Biochimie, [online] 58(8), p.XVIII. Available at: <http://sciencedirect.com/science/article/abs/pii/S0300908476802775?via%3Dihub> [Accessed 16 March 2022].
19. Reuter, P., 2000. Taschenwörterbuch der Biochemie / Pocket Dictionary of Biochemistry. [online] Available at: <https://link.springer.com/book/10.1007/978-3-0348-5081-0> [Accessed 16 March 2022].
20. Šesták, Z., 2000. Reuter, B.: Taschenwörterbuch der Biochemie. Deutsch - Englisch, Englisch - Deutsch. Pocket Dictionary of Biochemistry. English - German, German - English. Photosynthetica, [online] 38(4), pp.538-538. Available at: <https://ps.ueb.cas.cz/artkey/phs-200004-0013_reuter-b-taschenw-rterbuch-der-biochemie-deutsch-englisch-englisch-deutsch-pocket-dictionary-of-bioch.php> [Accessed 16 March 2022].

21. Daley, G., Kanowski, D. and Price, L., 2022. An interesting case in clinical biochemistry. Pathology, [online] 54, p.S4. Available at: <https://www.pathologyjournal.rcpa.edu.au/article/S0031-3025(21)00557-2/fulltext> [Accessed 16 March 2022].
22. Doran, H., 2022. Enterprising biochemistry. The Biochemist, [online] 44(1), pp.1-1. Available at: <https://portlandpress.com/biochemist/article/44/1/1/230686/Enterprising-biochemistry> [Accessed 16 March 2022].
23. Glenz, W., 2021. Dictionary. A Glossary of Plastics Terminology in 8 Languages, [online] pp.2-317. Available at: <https://www.hanser-elibrary.com/doi/10.3139/9781569908600.001> [Accessed 16 March 2022].
24. Harvey, J., 2022. Erythrocyte Biochemistry. Schalm's Veterinary Hematology, [online] pp.166-171. Available at: <https://onlinelibrary.wiley.com/doi/10.1002/9781119500537.ch21> [Accessed 16 March 2022].
25. Konwar, B., 2022. Biochemistry of Biosurfactants. Bacterial Biosurfactants, [online] pp.37-48. Available at: <https://www.taylorfrancis.com/chapters/mono/10.1201/9781003188131-4/biochemistry-biosurfactants-bolin-kumar-konwar> [Accessed 16 March 2022].
26. Konwar, B., 2022. Biochemistry of Biosurfactants. Bacterial Biosurfactants, [online] pp.37-48. Available at: <https://www.taylorfrancis.com/chapters/mono/10.1201/9781003188131-4/biochemistry-biosurfactants-bolin-kumar-konwar> [Accessed 16 March 2022].
27. McGhee, M., Saxelby, C. and McKay, N., 2021. Biochemistry. A Guide to Laboratory Investigations, [online] pp.68-105. Available at: <https://www.taylorfrancis.com/chapters/mono/10.1201/9781003049685-5/biochemistry-michael-mcghee-caroline-saxelby-niall-mckay> [Accessed 16 March 2022].
28. Müntz, K., 1992. Buchbesprechung. Biochemie und Physiologie der Pflanzen, [online] 188(1), p.32. Available at: <https://www.sciencedirect.com/science/article/abs/pii/S0015379611802568?via%3Dihub> [Accessed 16 March 2022].
29. Nakayama, T., 2022. Biochemistry and regulation of aurone biosynthesis. Bioscience, Biotechnology, and Biochemistry, [online] Available at: <https://academic.oup.com/bbb/advance-article-abstract/

doi/10.1093/bbb/zbac034/6544701?redirectedFrom=fulltext&login=false> [Accessed 16 March 2022].

30. Rajesh Prasad Rastogi, 2021. Ecophysiology and Biochemistry of Cyanobacteria. [online] Available at: <https://link.springer.com/book/10.1007/978-981-16-4873-1> [Accessed 16 March 2022].
31. Advances in Biochemistry in Health and Disease, 2022. Biochemistry of Apoptosis and Autophagy. [online] Available at: <https://link.springer.com/book/10.1007/978-3-030-78799-8> [Accessed 16 March 2022].
32. Daley, G., Kanowski, D. and Price, L., 2022. An interesting case in clinical biochemistry. Pathology, [online] 54, p.S4. Available at: <https://www.pathologyjournal.rcpa.edu.au/article/S0031-3025(21)00557-2/fulltext> [Accessed 16 March 2022].
33. KomparaLukančič, M., 2021. Jezik in turizem, Language and Tourism, Sprache und Tourismus. [online] Available at: <https://press.um.si/index.php/ump/catalog/book/635> [Accessed 16 March 2022].
34. Lawrence, R., 2022. Biochemistry of Human Milk. Breastfeeding, [online] pp.93-144. Available at: <https://www.sciencedirect.com/science/article/pii/B9780323680134000043?via%3Dihub> [Accessed 16 March 2022].
35. Leane, J., 2022. Wiradjuri Dictionary. Commonwealth Essays and Studies, [online] 44(2). Available at: <https://journals.openedition.org/ces/11133> [Accessed 16 March 2022].
36. Liu, Y., Liu, J., Long, Z. and Zhu, C., 2021. Tensor Dictionary Learning. Tensor Computation for Data Analysis, [online] pp.59-91. Available at: <https://link.springer.com/chapter/10.1007/978-3-030-74386-4_3> [Accessed 16 March 2022].
37. Manning, S., 2022. Microalgal lipids: biochemistry and biotechnology. Current Opinion in Biotechnology, [online] 74, pp.1-7. Available at: <https://www.sciencedirect.com/science/article/abs/pii/S0958166921002056?via%3Dihub> [Accessed 16 March 2022].
38. Melentyeva, Y., 2021. "Reading. The encyclopedic dictionary." Scientific and Technical Libraries, [online] (11), pp.147-152. Available at: <https://ntb.gpntb.ru/jour/article/view/867> [Accessed 16 March 2022].
39. Ochs, R., 2021. Biochemistry. [online] Available at: <https://www.taylorfrancis.com/books/mono/10.1201/9781003029649/biochemistry-raymond-ochs> [Accessed 16 March 2022].

40. Wang, Q., Li, Y. and Liu, P., 2022. Physiology and biochemistry of myxomycetes. Myxomycetes, [online] pp.231-267. Available at: <https://www.sciencedirect.com/science/article/pii/B9780128242810000026?via%3Dihub> [Accessed 16 March 2022].

INDEX

I

J

L

M

N

O

P

Q

R

Y

Z